ants, BIKES, & CLOCKS

ants, BIKES, & CLOCKS

PROBLEM SOLVING FOR UNDERGRADUATES

William Briggs
University of Colorado at Denver
Denver, Colorado

Society for Industrial and Applied Mathematics
Philadelphia

MATLAB® is a registered trademark of The MathWorks, Inc. For MATLAB product information please contact The MathWorks, Inc., 3 Apple Hill Drive, Natick, MA 01760-2098 USA, 508-647-7000, Fax: 508-647-7101, info@mathworks.com, www.mathworks.com/.

Library of Congress Cataloging-in-Publication Data

Briggs, William L.
  Ants, bikes, and clocks : problem solving for undergraduates / William Briggs.
    p. cm.
  Includes bibliographical references and index.
  ISBN 0-89871-574-1 (pbk.)
    1. Problem solving. I. Title.

QA63.B72 2005
510–dc22   2004058720

 is a registered trademark.

Contents

Chapter 1
Introduction

*Skills are to mathematics what scales are to music
or spelling is to writing. The objective of learning
is to write, to play music, or to solve problems
—not just to master skills.*

— Everybody Counts

For students of mathematics, science, and engineering, the name of the game is *problem solving*. Whether it's proving a theorem, writing a computer program, designing a statistical experiment, or solving for the stresses in a bridge, the essential challenge is problem solving.

Mathematical problem solving has been the subject of countless books, magazines, anthologies, and, more recently, web sites. It has been promoted and enjoyed by diverse audiences that include students, teachers, and recreational aficionados. Given the immense scope of problem solving, it's clear that one book cannot possibly do the subject justice.

To explain why this book may be different than many others, let's attempt a brief taxonomy of mathematical problem solving. Greatly simplifying the matter, we might identify the following categories of problems that mathematics students typically encounter (the list has no special order, and the categories certainly overlap):

1. **Recreational problems** are concise intellectual challenges often associated with puzzle connoisseurs such as Sam Loyd [22, 23] and Martin Gardner [10, 8, 9, 11]. These problems may or may not be mathematical in nature, but they generally require keen critical thinking and ingenious strategies. Books of many flavors and difficulties are devoted to recreational problems, and any serious problem solver should be familiar with them. If nothing else, they provide excellent mind calisthenics and occasionally come in handy as party tricks.

2. **Contest problems** are precisely formulated mathematical problems that often appear in formal exams and competitions such as the American High School Math Exam (AHSME), the USA Mathematical Olympiad (USAMO), the

It has been said that the most ancient of all puzzles is the Riddle of the Sphinx: What creature walks on four feet in the morning, two feet at noon, and three feet in the evening? Many people failed to answer the riddle until Oedipus, the future King of Thebes, gave the intended answer: Man crawls on all fours as a baby, walks upright as an adult, and uses a staff in his old age.

International Mathematical Olympiad (IMO), and the Putnam Exam. Such contest problems generally require a fair amount of mathematical background and sophistication.

Information about the Putnam exam can be found at http://math.scu.edu/putnam/.

3. **Logic problems** are generally qualitative in nature and often take the form of a story. Their solution requires organized thinking and often formal logic. Collections of logic problems abound; in fact, monthly magazines of logic problems can be found in supermarkets. These problems provide excellent thinking exercises and are often used as training for standardized exams such as the GRE, LSAT, and MCAT.

4. **Modeling** or **story problems** are quantitative problems that are posed in a realistic context. A key distinction of these problems is that they are *not* posed explicitly as mathematical problems. For this reason, their solution requires an essential preliminary step that may be the crux of the solution. That step, often called *modeling*, is to transform the stated problem from words into mathematics. Having formulated the problem in mathematical terms, it must still be solved! The only formal exam that emphasizes such problems is the Mathematical Contest in Modeling.

More information about the Mathematical Contest in Modeling can be found at http://www.comap.com/undergraduate/contests/mcm/.

5. The previous four categories require analytical techniques—traditional methods carried out with pencil and paper (and brain). Twenty-five years ago, the list would have ended here. However, we must now acknowledge that there is another tool in the problem-solving arsenal: the computer. With its numerical, graphical, and symbolic capabilities, the computer is the laboratory of mathematics; it is a tool of exploration and discovery. Without diminishing the role of analytical methods, it is fair to say that fluency with a programming language (for example, C++ or Java) or a mathematical environment (for example, MATLAB, Maple, Derive, or Mathematica) is an essential skill for all mathematics students. This observation leads to another problem category: **Computational problems** are those problems for which the powers of a computer are used for insight, exploration, or solution.

Even if oversimplified, these problem categories certainly demonstrate that problem solving is a diverse and complex enterprise. In choosing the focus of this book, choices had to be made. Because problems posed in a realistic context are common and important, and because solving such problems is a valuable skill, *this book highlights modeling or story problems.* This is not to say that the other categories are neglected, but the emphasis of the book is decidedly on the following two-step process:

- transforming a problem in context (a story problem) into a mathematical problem, and

- using both analytical and computational tools to solve the resulting mathematical problem.

If you have gotten this far, you have read about 1% of this book! Before going any further you should know a few honest facts. Most importantly, this book (or any problem-solving book) will not give you a universal formula for problem solving;

such a formula just doesn't exist. However, this book (and many others) can help you become a better problem solver.

For everyone, problem solving power comes with practice. Through practice, you see similar problems and patterns reappear. Through practice, you master techniques and variations on those techniques. And, perhaps most important of all, through practice, you gain confidence.

It's easy to say that practice leads to mastery, but practice is not always easy. A pianist cannot practice enough to improve unless she finds some enjoyment in practice. Analogously, to improve at problem solving, you really must find some enjoyment in mathematics and problem solving. And just as a competitive cyclist cannot seriously train unless he has a desire to win, proficiency in problem solving will be hard-earned without the desire to be a better problem solver. Hopefully, this book can provide enjoyment and instill desire.

A word about answers and solutions is in order. A lone *answer*, which is typically a numerical result such as $\frac{\pi}{4}$ or 8.23 miles, is never acceptable. It must always be accompanied by a complete *solution*, which is a full account of how you arrived at the answer. One goal of the book is to improve mathematical communication, both written and verbal. With mathematical proofs, the standards of exposition are fairly clear. With the more open-ended story problems that appear in this book, the rules may not be so evident. However, here are some guidelines.

Solutions must be compelling and convincing to others who have never seen the problem, but who have the mathematical background needed to understand the solution. Graphs, tables, and figures can often make a solution much more digestible. You should write a solution so that if *you* were to read it five years from now, you could make sense of your own writing.

Throughout the book, you are encouraged to focus as much on the *process* of problem solving as on the final answer. It often helps to step back and watch yourself in the act of problem solving. Are certain environments more conducive to successful problem solving? For example, do you do better on a bus, in the shower, surrounded by silence, or inside a set of vibrating headphones? Are certain times of day more effective? Are you more successful after a big pasta dinner or on an empty stomach? What strategies worked? Were there breakthrough moments when you experienced a key *Aha!* or *Eureka!* moment? Does working in groups help or hinder your problem solving? Try to follow your efforts, both the victories and the frustrations. You will learn a lot about problem solving by watching yourself in the act.

Having said that the end result isn't everything, it is always rewarding to know when you devise a correct solution. In this book, you will find hints and answers for most problems at the end of each chapter. Solutions to selected odd problems (marked with a ◇) appear at the back of the book. *Remember that very little is gained by reading the solution to a problem before seriously attempting to solve it.*

Finally, mathematical problems are like folklore: The origins of many problems are lost in the shadows of time. Problems fall into obscurity and are rediscovered; story lines change as problems are passed on. As an inveterate collector, I have regrettably forgotten where I first saw or heard many problems, and even if I could remember the source, it might not be the primary source. I have done my best to give credit for problems used in this book by citing *my* primary source. I apologize (and would like to be informed) if I have failed to recognize the creator of a good problem.

Creativity can solve almost any problem. The creative act, the defeat of habit by originality, overcomes everything.
— George Lois

The Lord's Prayer is 66 words, the Gettysburg Address is 286 words, and there are 1,322 words in the Declaration of Independence. Yet, government regulations on the sale of cabbage total 26,911 words.
— David McIntosh

Acknowledgments: I am grateful to four years of problem solving students at the University of Colorado at Denver who patiently worked through early drafts of this book. Thanks also to Mike Kawai of the Mathematics Department at CU-Denver for doing an accuracy check of many of the solutions. John Rogosich of Techsetters, Inc., provided valuable advice with the LaTeX preparation of the manuscript; I am grateful for his assistance. Thanks also to Linda Thiel, Donna Witzleben, Simon Dickey, and Andrea Missias at SIAM, who offered editorial guidance that made the book much better than it would have been otherwise. Finally, this book is dedicated to my father, who solved many problems in his lifetime.

1.1 Exercises

1.1. Read carefully! Read these questions carefully, pay attention to meanings of words, and be open-minded!

(i) Anna had six apples and ate all but four of them. How many apples were left?

(ii) If there are 12 one-cent stamps in a dozen, how many two-cent stamps are there in a dozen?

(iii) The butcher is six foot, four inches tall and wears size 14 shoes. What does he weigh?

(iv) If Mr. Kerry's rooster laid an egg in Mr. Bush's yard, who owns the egg?

(v) Is it legal to marry your widow's sister?

(vi) A taxi traveled from the hotel to the airport at an average speed of 40 miles per hour, and the trip took an hour and 20 minutes. It then traveled from the airport back to the hotel and again averaged 40 miles per hour. This time the trip took 80 minutes. Explain. [10]

(vii) A lady did not have her driver's license with her when she failed to stop at a stop sign and then went three blocks down a one-way street the wrong way. A policeman saw her, but he did not stop her. Explain.

1.2. Mixed fruit. One of three boxes contains apples; another box contains oranges; and another box contains a mixture of apples and oranges. The boxes are labelled APPLES, ORANGES, and APPLES AND ORANGES, but each label is incorrect. Can you select one fruit from only one box and determine the correct labels? Explain.

1.3. Fathers and sons. "Brothers and sisters I have none, but that man's father is my father's son." Who is *that man*?

1.4. Lost brothers. I am the brother of the blind fiddler, but brothers I have none. How can this be?

1.5. Small town haircuts. A visitor arrived in a small Nevada town in need of a haircut. He discovered that there were exactly two barbers in town. One was well groomed with splendidly cut hair, the other was unkempt with an unattractive hair cut. Which barber should the visitor patronize?

1.6. Revolving coins. Two quarters rest next to each other on a table. One coin is held fixed while the second coin is rolled around the edge of the first coin with no slipping. When the moving coin returns to its original position, how many times has it revolved?

1.7. Mixed apples. Three kinds of apples are all mixed up in a basket. How many apples must you draw (without looking) from the basket to be sure of getting at least two of one kind?

1.8. Socks in the dark. Suppose you have 40 blue socks and 40 brown socks in a drawer. How many socks must you take from the drawer (without looking) to be sure of getting (i) a pair of the same color, and (ii) a pair with different colors?

1.9. ◇ Birthday dilemma. Reuben says, "Two days ago I was 20 years old. Later next year I will be 23 years old." Is this possible? Explain.

1.10. Banquet counting. There were 100 basketball and football players at a sports banquet. Given any two athletes, at least one was a basketball player. If at least one athlete was a football player, how many football players were at the banquet? [39]

1.11. ◇ A rising tide. A rope ladder hanging over the side of a boat has rungs one foot apart. Ten rungs are showing. If the tide rises five feet, how many rungs will be showing?

1.12. Chocolate demographics. Half of all people are chocolate eaters and half of all people are women. (i) Does it follow that $\frac{1}{2} \times \frac{1}{2} = \frac{1}{4}$ of all people are female chocolate eaters? (ii) Does it follow that half of all men are chocolate eaters? Explain.

1.13. ◇ **Family chess**. A woman, her older brother, her son, and her daughter are chess players. The worst player's twin, who is one of the four players, and the best player are of opposite sex. The worst player and the best player have the same age. If this is possible, who is the worst player? [24]

1.14. Subway dating. A Manhattan fellow had a girlfriend in the Bronx and a girlfriend in Brooklyn. He decided which girlfriend to visit by arriving randomly at the train station and taking the first of the Bronx or Brooklyn trains that arrived. The trains to Brooklyn and the Bronx *each* arrived regularly every 10 minutes. Not long after he began his scheme the man's Bronx girlfriend left him because he rarely visited. Give a (logical) explanation. [8]

1.15. ◇ **Chiming clock**. If a clock takes 5 seconds to strike 5:00 (chiming five times), how long does it take to strike 10:00 (chiming ten times)? Assume that the chimes occur instantaneously.

1.16. Mislabelled babies. One day in the maternity ward, the name tags for four girl babies became mixed up. (i) In how many different ways could two of the babies be tagged correctly and two of the babies be tagged incorrectly? (ii) In how many different ways could three of the babies be tagged correctly and one baby be tagged incorrectly?

1.17. ◇ **Card trick**. Alex says to you, "I'll bet you any amount of money that if I shuffle this deck of cards, there will always be as many red cards in the first half of the deck as there are black cards in the second half of the deck." Should you accept his bet? [*American Mathematical Monthly*, 26, January 1953, p.167]

1.18. Authors on a train. Six authors are riding in a train compartment. They are sitting three on a side, facing each other, with each author directly opposite one other author. Their names are Black, Brown, Red, Green, Pink, and White, and they are (not in the same order) an essayist, a historian, a humorist, a novelist, a playwright, and a poet. Each author has written a book which another occupant of the compartment is reading. Use the following clues to identify the six authors by where they sit, what they write, and what they read.

(i) Black is reading essays.

(ii) Red is reading a book by the author sitting directly opposite him.

(iii) Brown, the essayist, and the humorist sit on the same side, with Brown in the middle.

(iv) Pink is sitting next to the playwright.

(v) The essayist is sitting directly opposite the historian.

(vi) Green is reading plays.

(vii) Brown is the novelist's brother-in-law.

(viii) Black, who is in a corner seat, neither reads nor writes history.

(ix) Green is sitting directly opposite the novelist.

(x) Pink is reading a humor book.

(xi) White never reads poetry.

1.19. ◇ **Truth and lies**. An explorer lands on a strange island with two towns. All inhabitants of Truth always tell the truth, and all inhabitants of Lies always lie. The explorer asks the first islander (A) which town she comes from, and she gives an answer that the explorer cannot understand. The second islander (B) says, "A said she is from Lies." The third islander (C) says to B, "you are a liar!" From which town is C?

1.20. Sons and daughters. Suppose that each daughter in your family has the same number of brothers as she has sisters, and each son in your family has twice as many sisters as he has brothers. How many sons and daughters are in the family?

1.21. ◇ **Book orders**. Five books of five different colors are placed on a shelf. The orange book is between the gray and pink books, and these three books are consecutive. The gold book is not first on the shelf and the pink book is not last. The brown book is separated from the pink book by two books. If the gold book is not next to the brown book, what is the complete order of the five books? [39]

1.22. Scale calibration. The zero point on a bathroom scale is set incorrectly, but otherwise the scale works fine. It shows 60 kilograms when Dan stands on the scale, 50 kilograms when Sarah stands on the scale, but 105 kilograms when Dan and Sarah both stand on the scale. Does the scale read too high or too low? Explain. [40]

1.23. ◇ **A very old puzzler**. Three guests checked in at a hotel and paid $30 for their room. A while later, the desk clerk realized that the room should have cost only $25, so she gave the bellboy $5 to return to the three guests. The bellboy realized that $5 couldn't be divided evenly among the three guests so he kept $2 and gave $1 to each of the guests. It seems that the three guests have now each spent $9 for the room (that makes $27) and the bell boy has $2, for a total of $29. Where is the missing dollar?

1.2 Hints and Answers

1.1. (i) ANSWER: Four apples are left.

(i) ANSWER: There are 12 two-cent stamps in a dozen.

(ii) ANSWER: The butcher weighs meat.

(iii) ANSWER: Roosters do not lay eggs.

(iv) ANSWER: If your wife is a widow, you would not be alive.

(v) ANSWER: An hour and 20 minutes is 80 minutes.

(vi) ANSWER: The lady was walking.

1.2. HINT: Use trial and error. See what happens if you use each box for selecting the single fruit.
ANSWER: The fruit must be selected from the box labeled APPLES AND ORANGES.

1.3. HINT: Draw a diagram showing the speaker and *that man.*
ANSWER: *That man* is the son of the speaker.

1.4. HINT: Avoid gender bias!
ANSWER: The blind fiddler is a woman.

1.5. HINT: Who cuts the hair of the barbers?
ANSWER: Go to the unkempt barber, who must have given the well-groomed barber his haircut.

1.6. HINT: Try it!
ANSWER: The moving quarter revolves two full revolutions.

1.7. HINT: Find the worst possible case.
ANSWER: You must draw at most four apples.

1.8. HINT: Find the worst possible cases.
ANSWER: (i) You must draw three socks. (ii) You must draw 41 socks.

1.9. HINT: Draw a time line and look at special days of the year.
ANSWER: Yes, it is possible. Reuben was born on December 31 and spoke on January 1.

1.10. HINT: Interpret the sentence, "given any two athletes, at least one was a basketball player."
ANSWER: There are 99 basketball players and 1 football player.

1.11. HINT: A rising tide lifts all ships.
ANSWER: Ten rungs will be showing.

1.12. HINT: Being a man or a woman is independent of preference for chocolate.
ANSWER: (i) No, it does not follow. (ii) No, it does not follow.

1.13. HINT: Look for consistency of the facts.
ANSWER: The situation is not possible.

1.14. HINT: What are the various ways in which each train might arrive regularly every ten minutes?
ANSWER: For example, the Bronx train arrives at 12:00, 12:10, 12:20, ... and the Brooklyn train arrives at 12:09, 12:19, 12:29, ...

1.15. HINT: Consider the time between chimes.
ANSWER: It takes $\frac{45}{4} = 11.25$ seconds to strike 10:00.

1.16. HINT: How many different pairs of correctly named babies are possible?
ANSWER: (i) There are six ways that two of the four babies can be correctly tagged. (ii) There are no ways in which three of the four babies can be correctly tagged.

1.17. HINT: Consider some special cases.
ANSWER: Alex is correct; you should not accept his bet.

1.18. HINT: Organize the given information using a table or seating chart.
ANSWER: Here is the seating chart with each writer's choice of reading. Is it unique?

White, essayist reads history	Brown, poet reads novels	Green, humorist reads plays
Pink, historian reads humor	Red, playwright reads poetry	Black, novelist reads essays

1.19. HINT: Regardless of what B and C said, what must A have said?
ANSWER: C is from Truth.

1.20. HINT: Find the two conditions for the two unknowns.
ANSWER: There are three sons and four daughters.

1.21. HINT: The facts that the brown book is separated from the pink book by two books and that the pink book is not last lead to three cases.
ANSWER: The order is brown, gray, orange, pink, gold.

1.22. HINT: Suppose Dan is on the scale when Sarah steps on the scale. How much does the scale reading increase?
ANSWER: The scale reads five kilograms too high.

1.23. HINT: Follow the dollars and do not be deceived by the (faulty) arithmetic.

Chapter 2

A Problem-Solving Framework

If you don't know where you're going, you'll probably end up someplace else.

— Yogi Berra

When it comes to mathematical problem solving, there are innate abilities and acquired skills. Let's agree that changing one's innate abilities is itself an unsolvable problem! That leaves the challenge of improving one's acquired skills. This *is* a solvable problem, or at least one on which progress can be made.

A substantial literature exists on the theoretical side of mathematical problem solving [34, 35]. Much of it involves theories of learning, cognitive psychology, and other esoteric disciplines. It's worthwhile knowing about this research, and it may eventually lead to some practical understanding of problem solving. However, we will take a different approach that is practical from the start: We will solve problems.

One goal of this book is to acquaint you with yourself as a problem solver. Through plenty of practice, this book encourages you to learn how you approach problems, to determine what conditions and strategies work best for you, and to strengthen your problem-solving skills.

> A great discovery solves a great problem but there is a grain of discovery in the solution of every problem.
> — George Pólya

As mentioned in the Introduction, no single prescription or formula works for all problems and all problem solvers. However, this chapter presents a general problem-solving framework, followed in the next chapter by many practical strategies. Taken altogether, they provide a fairly powerful approach to problem solving.

2.1 Pólya's Method

In 1945, the mathematician George Pólya wrote a small book called *How To Solve It* (see [29]). Since then it has become one of the most widely read mathematics books, and for good reason. It is a practical guide to mathematical problem solving, filled with hints and examples. Clearly, it is required reading for all problem solvers.

The foundation of Pólya's book is a four-step procedure that can be used to organize the problem-solving process. It is not a specific prescription that works for all problems, but it is a useful set of guidelines. You can think of the four-step procedure as a *framework*; it is like a four-room house. The procedure tells you to visit each room; however, it does not tell you *exactly* what to do in each room. Here is Pólya's four-step procedure with specific annotations for each step.

George Pólya was born in Hungary in 1887 and came to the United States during World War II. He spent most of the rest of his life at Stanford University, where he continued to be a prolific mathematician. He died in Palo Alto, California in 1985 at the age of 97.

Pólya's Method

Step 1: Understand the problem. The first step in problem solving is to determine where you are going. Be sure that you understand what the problem is asking.

- Read the problem carefully! If it helps, read it aloud.
- Record the quantities and conditions that are given (often called the *data* for the problem).
- Identify the unknowns. Exactly what is to be determined?
- Draw a picture or diagram to help you organize the information and visualize the problem.
- If possible, restate the problem in different ways to clarify it.

Step 2: Plan a strategy for solving the problem. Once you understand the problem, the next step is to decide how to go about solving it. This step is the most difficult; it requires creativity, organization, and experience.

- Try to think of a similar or related problem.
- Map out your strategy with a flow chart or diagram.
- Identify the appropriate analytical or computational tools needed for the solution.
- Apply the hints given in Chapter 3.

Strategy is a style of thinking, a conscious and deliberate process, an intensive implementation system, the science of insuring future success.
— Pete Johnson

Step 3: Execute your strategy, and revise it if necessary. After devising a strategy, the next step is to carry it out.

- Keep an organized written record of your work, which will be helpful if revisions are needed.
- Double-check each step so that you do not propagate errors to the end of the solution.
- Assess your strategy as you work; if you find a flaw, return to Step 2 and revise your strategy.

Step 4: Check and interpret your result. It's tempting to stop after Step 3; however, the final step may be the most important.

- Be sure that your result makes sense; for example, check that it has the expected units and that numerical values are sensible.
- Recheck your calculations, or find an independent way of checking the result.
- Check the consistency of the result by considering special or limiting cases.
- Write the solution clearly and concisely.

Pólya's method can be summarized as *Understand, Plan, Execute, Check*. It is a good idea to follow the procedure as closely as possible while you are getting used to it. However, you may also discover that it eventually becomes part of your problem-solving mindset, much as you use a microwave or VCR without thinking about the individual steps. Pólya's method will be implicit in every example of this book, even if it is not written out explicitly for each solution.

> The mark of a good action is that it appears inevitable in retrospect.
> — R.L. Stevenson

Example 2.1 Passing boats. Two boats leave from opposite banks of a river at the same time and travel at constant but different speeds. They pass each other 700 yards from one bank and continue to the other side of the river, where they turn around. On their return trip the boats pass again—this time 400 yards from the opposite bank. How wide is the river? [originally posed by W.C. Rufus, *American Mathematical Monthly*, 47, February 1947]

Solution: Step 1: Understand the problem. After reading the problem—perhaps more than once, perhaps aloud—you should have an image of two boats shuttling back and forth across a river. The boats leave opposite banks at the same time, passing each other the first time 700 yards from one bank. They turn around and pass each other the second time 400 yards from the opposite bank. *A picture is highly recommended to summarize these facts*, and one is provided in Figure 2.1.

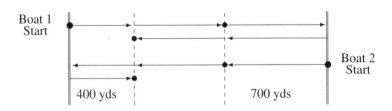

Figure 2.1. *A picture illustrating the passing of two boats greatly facilitates the solution of the problem.*

A key fact is that the boat speeds are constant but different. When working with constant speeds, the relationship

$$\text{distance} = \text{speed} \times \text{time elapsed}$$

can always be invoked.

Step 2: Plan a strategy for solving the problem. With an image of the problem in mind, we can plan. Here is the **key insight**: The time elapsed for the first leg, between the boats' departure and their first passing, is the same for both boats; furthermore, we know the location of the first passing point (700 yards from one bank). Note also that the time elapsed for the second leg, between the first and second passings, is also the same for both boats, and we know the location of the second passing point (400 yards from the opposite bank). We might anticipate that these two observations will lead to two relationships among the variables in the problem.

Thinking ahead, we will need to define the unknown quantities: Let v_1 and v_2 be the speeds of the boats, and let w be the width of the river. It appears that we will have only two relationships for three unknowns. However, it's worth remembering that often only the *ratio* of unknowns (such as v_1 and v_2) is needed, not the values of the unknowns themselves. Clearly, working with ratios reduces the number of unknowns.

Furious activity is no substitute for understanding.
— H.H. Williams

Step 3: Execute your strategy. Let's assume that Boat 1, with speed v_1, travels $w - 700$ yards on the first leg, which means that Boat 2, with speed v_2, travels 700 yards on the first leg. Because the travel times for the boats are the same on the first leg, we have

$$\text{time elapsed} = \frac{w - 700}{v_1} = \frac{700}{v_2}.$$

Because the travel times for the boats are the same on the second leg, we have (using Figure 2.1)

$$\text{time elapsed} = \frac{700 + w - 400}{v_1} = \frac{w - 700 + 400}{v_2}.$$

Both of these equations can be solved for the ratio v_1/v_2 to give

$$\frac{v_1}{v_2} = \frac{w - 700}{700} = \frac{300 + w}{w - 300}.$$

Much as we hoped in Step 2, the ratio of the speeds can be eliminated from the problem, and a single equation for the river width remains. The algebra is straightforward; the equation that must be solved is

$$w^2 - 1700w = 0,$$

which has solutions $w = 0$ and $w = 1700$ yards. Clearly, only the second solution is meaningful in this problem.

Step 4: Check and interpret your result. The solution $w = 1700$ is numerically reasonable, and it is easy to check that both relations are satisfied with $w = 1700$. We can also determine that the ratio of the boat speeds is $\frac{v_1}{v_2} = \frac{10}{7}$; the actual boat speeds cannot be found. Finally, Figure 2.1 is drawn assuming that the first passing point is nearer the right bank than the left bank. It should be verified that the solution is unchanged if this assumption is reversed.   □

Example 2.2 The three prisoners. Three prisoners know that the jailer has three white hats and two red hats. The jailer puts a hat on the head of each prisoner and says, "If you can deduce the color of your own hat, you will be freed." Each prisoner can see the hats of the other two prisoners but not his own. The first prisoner says, "I cannot tell the color of my hat." Then the second prisoner says, "I cannot tell the color of my hat." The third prisoner, who is blind, is able to determine the color of his hat and is freed. What is the color of the third prisoner's hat, and how did he know?

Solution: Step 1: Understand the problem. It is essential to visualize the situation clearly. Picture three prisoners, one of whom is blind, each with a hat on his head; the hats have been selected from a group of three white hats and two red hats. We must also assume that all three prisoners are thinking clearly and logically.

Step 2: Plan a strategy for solving the problem. There are $2^3 = 8$ possible ways that the prisoners could wear red and white hats (two choices for each of the three prisoners). It helps to make a table showing all eight possibilities. Then from the first two prisoners' statements, it should be possible to think like the blind prisoner and eliminate one hat color or the other.

Step 3: Execute your strategy. Table 2.1 shows the eight possible cases, together with the thoughts of the blind prisoner (R = red, W = white). Note that Case 1 cannot occur because the jailer has only two red hats.

I'm not smart, but I like to observe. Millions saw the apple fall, but Newton was the one who asked why.
— Bernard Baruch

Table 2.1. *The table shows the eight cases that arise in the prisoner problem* (R = *red*, W = *white*).

	Prisoner 1	Prisoner 2	Prisoner 3	Thoughts of Prisoner 3
1	R	R	R	Impossible; only two red hats
2	R	R	W	
3	R	W	R	Eliminated by Prisoner 2's statement
4	R	W	W	
5	W	R	R	Eliminated by Prisoner 1's statement
6	W	R	W	
7	W	W	R	Eliminated by Prisoner 2's statement
8	W	W	W	

Now put yourself in the mind of the blind prisoner. When the first prisoner speaks, he knows that Cases 1 and 5 can be eliminated, the latter because if the first prisoner saw two red hats, he would know that he had a white hat. At this point, everyone knows that Prisoner 2 and the blind prisoner cannot both have red hats. Therefore, if Prisoner 2 sees a red hat on the blind prisoner, Prisoner 2 knows he must have a white hat. Because Prisoner 2 cannot determine the color of his own hat, he must have seen a white hat on the blind prisoner. So Prisoner 2's statement tells the blind prisoner that Cases 3 and 7 can also be eliminated. The remaining options all require the blind prisoner to have a white hat. Thus, the blind prisoner knows that he has a white hat.

Step 4: Check and interpret your result. In this case, it would be wise to review the arguments that led to the elimination of Cases 3, 5, and 7. Having confirmed that they are valid, the conclusion stands. ☐

Hopefully, these two detailed examples give an adequate illustration of Pólya's method at work. The best way to learn is not by reading but by practicing. The following exercises provide an opportunity to use Pólya's method. For each problem, write out the steps of Pólya's method explicitly and comment on your problem-solving process.

2.2 Exercises

2.1. ◇ **Balls in a can**. Three tennis balls fit exactly (no room to spare) in a cylindrical can. Which is greater, the circumference of the can or the height of the can?

2.2. Fish anatomy. The length of the tail of a fish equals the length of its head plus a quarter the length of its body. Its body length is three-quarters of its total length. Its head is 4 inches long. What is the total length of the fish? [14]

2.3. ◇ **Socks**. A man who owns only 12 brown and 8 black socks puts all of his socks into a drawer without pairing them. (i) If he selects socks from the drawer at random, what is the smallest number of socks that he must select to be *certain* of having a pair of black socks? (ii) What is the smallest number of socks that he must select to be *certain* of having a pair of socks of either color? [1]

2.4. Dividing pennies. Alice takes one-third of the pennies from a large jar. Then Bret takes one-third of the remaining pennies from the jar. Finally, Carla takes one-third of the remaining pennies from the jar, leaving 40 pennies in the jar. How many pennies were in the jar at the start?

2.5. ◇ **Counting pages**. To number the pages of a book, a printer used 2989 digits, beginning at page 1. How many pages does the book have? [29]

2.6. Water lily. When the stem of a water lily is vertical, the blossom is 10 centimeters above the surface of the lake. Pulling the lily to one side, keeping the stem straight, the blossom touches the water at a point 21 centimeters from where the stem formerly cut the surface. How deep is the water in the lake? (attributed to Henry Wadsworth Longfellow in [10])

2.7. ◇ **Old age**. An ancient Greek was said to have lived one-fourth of his life as a boy, one-fifth as a youth, one-third as a man, and spent the last 13 years as an old man. How old was he when he died?

2.8. Coffee and milk. One morning each member of Angela's family drank an eight-ounce cup of coffee and milk, with the (nonzero) amounts of coffee and milk varying from cup to cup. Angela drank a quarter of the total amount of milk and a sixth of the total amount of coffee. What is the least number of people in the family?

2.9. ◇ **Apple sale**. When Stuart bought his usual 50 cents worth of apples, he found that he was able to buy 5 more apples than usual. Noting that the price of apples decreased by 10 cents per dozen, what is the new price per dozen? [17]

2.10. Spilled juice. Mort spilled half the contents of a pitcher of orange juice. He then filled the half-full pitcher with water. He poured himself a glass of the (diluted) orange juice, leaving the pitcher three-fourths full. Realizing the orange juice was too diluted, he filled the pitcher to the top with double strength orange juice. Compare the concentration of the original and final juices.

2.11. ◇ **Two bad clocks**. Of two clocks next to each other, one runs 5 minutes per hour fast and the other runs 5 minutes per hour slow. If the clocks start showing the same time, how long before they are one hour apart?

2.12. Comparing salaries. Bob takes a job in which he receives a starting salary of $3000 per year, with raises of $600 per year. Sue takes a job in which she receives a starting salary of $1500 per half year, with raises of $300 every six months. Which person is better paid? Explain. [14]

2.13. ◇ **Rain and shine**. During a vacation, it rained on 13 days, but when it rained in the morning, the afternoon was sunny, and every rainy afternoon was preceded by a sunny morning. There were 11 sunny mornings and 12 sunny afternoons. How long was the vacation? [14]

2.14. Handshakes. In a room with 23 people, at least one person has not shaken hands with everyone else in the room. What is the maximum number of people in the room who could have shaken hands with everyone else? [24]

2.15. ◇ **Passing bikes**. Two towns, A and B, are connected by a road. At sunrise, Pat begins biking from A to B, and Dana begins biking from B to A. They ride at constant (but different) speeds, and they pass each other at noon. Pat reaches B at 5:00 p.m., while Dana reaches A at 8:00 p.m. When was the sunrise? [43]

2.16. Puzzled. A jigsaw puzzle contains 500 pieces. A *section* of the puzzle is a set of one or more pieces that have been connected to each other. A *move* consists of joining two sections. What is the smallest number of moves in which the puzzle can be assembled?

2.17. Designing functions: ages, mileage, headstarts. In the following exercises, find a function that describes the given situation. Graph the function, and give the domain (values of the independent variable) that makes sense for the problem.

(i) ◇ Find a function $y = f(x)$ such that if A is a fraction x greater than B, then B is a fraction y less than A.

(ii) Find a function $y = f(x)$ such that if you are married at age 25, then at age x you have been married a fraction y of your life.

(iii) ◇ Find a function $y = f(x)$ such that if you have a brother two years older than you are, then at age x your age is a fraction y of your brother's age.

(iv) Find a function $y = f(x)$ such that if you have a sister two years younger than you are, then at age x you are a fraction y older than your sister.

(v) ◇ Find a function $y = f(x)$ such that if you can arrive at your destination by averaging 60 miles per hour, and halfway to your destination you discover that you have been averaging x miles per hour, then you must average at least y miles per hour for the remainder of the trip in order to arrive on time.

(vi) Find a function $y = f(x)$ such that if your car gets x miles per gallon and gasoline costs $1.60 per gallon, then $$y$ is the cost of taking a 350-mile trip.

(vii) ◇ Find a function $y = f(x)$ such that if your car gets 32 miles per gallon and gasoline costs $1.60 per gallon, then $$y$ is the cost of taking an x-mile trip.

(viii) Find a function $y = f(x)$ such that if your car gets 32 miles per gallon and gasoline costs $$x$ per gallon, then $100 is the cost of taking a y-mile trip.

(ix) ◇ Find a function $y = f(x)$ such that if A, averaging 30 miles per hour, has an x-hour headstart over B, averaging 40 miles per hour, then it takes B y hours to catch A.

(x) Find a function $y = f(x)$ such that if A, averaging x miles per hour, has a 0.6-hour headstart over B, averaging 40 miles per hour, then it takes B y hours to catch A.

(xi) ◇ Find a function $y = f(x)$ such that if A, averaging x miles per hour, has a 0.6-hour headstart over B, averaging y miles per hour, then it takes B 3 hours to catch A.

(xii) Find a function $y = f(x)$ such that if A, averaging y miles per hour, has an x-hour headstart over B, averaging 40 miles per hour, then it takes B 2.5 hours to catch A.

2.18. Designing functions: dilution problems. In the following exercises, find a function that describes the given situation. Graph the function and give the domain (values of the independent variable) that makes sense for the problem; specifically, state the values of x for which the given dilution is possible. Percentage concentrations are by volume and have values between 0 and 100.

(i) ◇ Find a function $y = f(x)$ such that when x liters are removed from 10 liters of a 40 percent alcohol solution and replaced by x liters of a y percent solution, the result is a 65 percent alcohol solution.

(ii) Find a function $y = f(x)$ such that when 2 liters are removed from 10 liters of a 40 percent alcohol solution and replaced by 2 liters of a x percent solution, the result is a y percent alcohol solution.

(iii) ◇ Find a function $y = f(x)$ such that when 2 liters are removed from 10 liters of an x percent alcohol solution and replaced by 2 liters of a y percent solution, the result is a 50 percent alcohol solution.

(iv) Find a function $y = f(x)$ such that when x liters are removed from 10 liters of a 40 percent alcohol solution and replaced by x liters of a 90 percent solution, the result is a y percent alcohol solution.

2.19. ◇ **Passing exams**. Forty-one students each took an exam in algebra, biology, and chemistry.

- 12 students failed (at least) the algebra exam.
- 5 students failed (at least) the biology exam.
- 8 students failed (at least) the chemistry exam.
- 2 students failed (at least) both the algebra and the biology exam.
- 6 students failed (at least) both the algebra and the chemistry exam.
- 3 students failed (at least) both the biology and the chemistry exam.
- 1 student failed all three exams.

How many students passed all three exams? [20]

2.20. In the running. Sven placed exactly in the middle among all runners in a race. In 10th place, Dan placed lower than Sven, and Lars was in 16th place. How many runners were in the race? [40]

2.21. ◇ **Monkeys and coconuts**. Five men and a monkey were shipwrecked on a desert island, and they spent the first day gathering coconuts for food. They piled them up together and then went to sleep for the night. But when they were all asleep one man woke up, and he thought there might be a row about dividing the coconuts in the morning, so he decided to take his share. So he divided the coconuts into five piles. He had one coconut left over, and he gave that to the monkey, and he hid his pile and put the rest back together. By and by the next man woke up and did the same thing. And he had one left over, and he gave it to the monkey. And all five of the men did the same thing, one after the other; each one taking a fifth of the coconuts in the pile when he woke up, and each one having one left over for the monkey. And in the morning they divided what coconuts were left, and they came out in five equal shares. Of course each one must have known there were coconuts missing; but each one was as guilty as the others, so they didn't say anything. How many coconuts were there in the beginning? [*Saturday Evening Post*, October 9, 1926].

2.22. Open and shut case. Five hundred students arrived for the first day of school and found 500 numbered lockers, all of which were closed. The first student switched every locker (opened the closed lockers and closed the open lockers); the second student switched every other locker (lockers $2, 4, 6, \ldots$). The nth student switched every nth locker (lockers $n, 2n, 3n, \ldots$). The pattern continued up to the 500th student. Give a simple description of the open/closed pattern of the lockers in their final state. [*Pi Mu Epsilon Journal*, 1, April 1953]

2.23. ◇ **Shoe shuffle**. Ann had 240 shoes in a large box, 80 in each of black, brown, and white. All the shoes had the same size and style, and half were for the left foot and half were for the right foot. Over all distributions of shoes, what are the minimum and maximum number of pairs of shoes that Ann is always assured of finding in the box? (A pair is a right and left shoe of the same color.) [40]

2.24. Ladders in an alley. Two ladders of length a lean against opposite walls of an alley with their feet touching. One ladder extends h feet up the wall and makes a 75° angle with the ground. The other ladder extends k feet up the wall and makes a 45° angle with the ground. In terms of a, h, and/or k, how wide is the alley? [24]

2.3   Hints and Answers

2.1. HINT: Draw a picture and note the relationship between the radius of the balls and the radius of the can.
ANSWER: The circumference of the can is greater than the height of the can.

2.2. HINT: Find three equations for the lengths of the tail, body, and head.
ANSWER: The total length of the fish is 128 inches.

2.3. HINT: Consider the worst possible situations for getting a pair of black socks and a pair of either color.
ANSWER: (i) He must select 14 socks to be sure of getting a black pair. (ii) He must select three socks to be sure of getting a pair of either color.

2.4. HINT: Each person takes one-third of the pennies and leaves two-thirds of the pennies.
ANSWER: There were 135 pennies in the jar.

2.5. HINT: Count the number of pages that have one-, two-, three-, and four-digit page numbers.
ANSWER: The book has 1024 pages.

2.6. HINT: Draw a good picture and use the Pythagorean theorem.
ANSWER: The water depth is approximately 17.05 centimeters.

2.7. HINT: Let a be the man's age at death and describe his life in terms of its four stages.
ANSWER: The man was 60 years old when he died.

2.8. HINT: Let M and C be the total amount of milk and coffee, respectively, consumed by the family. Write equations for the amount of milk and coffee in Angela's drink and for the entire family.
ANSWER: There are five people in the family.

2.9. HINT: Relate the number of apples Stuart could buy before and after the price change.
ANSWER: The new price is 30 cents per dozen.

2.10. HINT: Let Q be the amount of juice in the pitcher at the beginning and compute the amount of juice at each stage.
ANSWER: The concentration at the end is $\frac{7}{8}$ the concentration at the beginning.

2.11. HINT: Compare the times shown by the clocks after $1, 2, 3, \ldots$ hours.
ANSWER: After six hours, the two clocks show times that are one hour apart.

2.12. HINT: Compare the salaries after $1, 2, 3, \ldots$ years.
ANSWER: Sue is better paid.

2.13. HINT: There are four different kinds of days: rain/sun in the morning/afternoon.
ANSWER: The vacation lasted 18 days.

2.14. HINT: Consider some special cases with $3, 4, 5, \ldots$ people.
ANSWER: At most 21 people could have shaken hands with everyone.

2.15. HINT: Let t be the number of hours before noon that the sun rises and let L be the distance between the towns.
ANSWER: The sun rose $2\sqrt{10} \doteq 6.32$ hours before noon.

2.16. HINT: Consider some special cases with $3, 4, 5, \ldots$ puzzle pieces.
ANSWER: The minimum number of moves needed to assemble the puzzle is 499.

2.17. HINT: In all of these problems, a good way to get started is to consider special cases with specific values for x and y.

(i) ANSWER: $y = f(x) = \frac{x}{1+x}$. Domain $\{x : x \geq 0\}$.
(ii) ANSWER: $y = f(x) = \frac{x-25}{x}$. Domain $\{x : x \geq 25\}$.
(iii) ANSWER: $y = f(x) = \frac{x}{x+2}$. Domain $\{x : x \geq 0\}$.

(iv) ANSWER: $y = f(x) = \frac{2}{x-2}$. Domain $\{x : x > 2\}$.

(v) ANSWER: $y = \frac{30x}{x-30}$. Domain $\{x : x > 30\}$.

(vi) ANSWER: $y = \frac{560}{x}$. Domain $\{x : x > 0\}$.

(vii) ANSWER: $y = 0.05x$. Domain $\{x : x \geq 0\}$.

(viii) ANSWER: $y = \frac{3200}{x}$. Domain $\{x : x > 0\}$.

(ix) ANSWER: $y = 3x$. Domain $\{x : x \geq 0\}$.

(x) ANSWER: $y = \frac{0.6x}{40-x}$. Domain $\{x : 0 \leq x < 40\}$.

(xi) ANSWER: $y = 1.2x$. Domain $\{x : x \geq 0\}$.

(xii) ANSWER: $y = \frac{40}{1+0.4x}$. Domain $\{x : x \geq 0\}$.

2.18. HINT: In all of these problems, a good way to get started is to consider special cases with specific values for x and y.

(i) ANSWER: $y = f(x) = 40 + \frac{250}{x}$. Domain $\left\{x : \frac{25}{6} \leq x \leq 10\right\}$.

(ii) ANSWER: $y = f(x) = 32 + \frac{x}{5}$. Domain $\{x : 0 \leq x \leq 100\}$.

(iii) ANSWER: $y = f(x) = 250 - 4x$. Domain $\left\{x : \frac{75}{2} \leq x \leq \frac{125}{2}\right\}$.

(iv) ANSWER: $y = f(x) = 40 + 5x$. Domain $\{x : 0 \leq x \leq 10\}$.

2.19. HINT: The students who failed at least one exam include the students who failed at least two or three exams. Similarly, the students who failed at least two exams include the students who failed all three exams.

ANSWER: Twenty-six students passed all three exams.

2.20. HINT: The faster the runner, the lower his place. The number of runners in the race is odd.

ANSWER: There were 17 runners in the race.

2.21. HINT: A short solution is subtle. So attack it head-on by writing equations for each step of the process.

ANSWER: The smallest possible number of coconuts is 3121.

2.22. HINT: Carry out the process for, say, 20 lockers, and look for a pattern.

ANSWER: At the end, the numbers of the open lockers are perfect squares $(1, 4, 9, 16, \ldots)$.

2.23. HINT: Read the problem carefully and understand exactly what is asked. Make a table showing all six types of shoes, and consider various distributions of shoes subject to the constraints.

ANSWER: Ann is assured of finding no more than or no less than 40 pairs of shoes.

2.24. HINT: Draw a good picture and use either geometry or trigonometry.

ANSWER: The width of the alley is h.

Chapter 3
Problem-Solving Strategies

When you have eliminated the impossible, whatever remains, however improbable, must be the truth.

— Sir Arthur C. Doyle, *The Sign of Four*

Using Pólya's method will provide guidance and help you stay organized. However, it doesn't provide the specific techniques or strategies required to reach a solution. In this section, we offer a dozen specific problem-solving hints and strategies. They should be kept in mind while working with Pólya's method.

Hint 1: More than one answer.

Mathematics abounds with problems that have more than one numerical answer. For example, the equation $x^2 = 16$ has solutions $x = 4$ and $x = -4$, and, without further conditions, both solutions are equally acceptable. Such *nonuniqueness* occurs because not enough information is given to distinguish among a variety of possibilities.

Slightly more exciting, imagine standing on the edge of a cliff 100 feet above the ground. Suppose you throw a rock upward at a speed of 60 feet per second. The height of the rock above the ground at any time $t \geq 0$ is given approximately by $y(t) = -16t^2 + 60t + 100$. To find when the rock hits the ground, we must solve the equation $y(t) = 0$. The quadratic formula quickly gives the two mathematical solutions, $t = -\frac{5}{4}$ seconds and $t = 5$ seconds. However, the context of the problem requires $t > 0$, because the rock hits the ground *after* it is thrown. In this case, only the second solution is physically acceptable.

Example 3.1 Traffic counters. A traffic counter is a cable stretched across a road, connected to a box at the side of the road. The device registers one count each time a pair of wheels on a single axle rolls over the cable. A normal automobile registers two counts: one for the front wheels and one for the rear wheels. A light truck with three axles (front wheels plus a double set of rear wheels) registers three counts. A heavy truck might register four counts. Suppose that during a one-hour period a particular traffic counter registers 35 counts on a street on which two-axle, three-axle, and four-axle vehicles are allowed. How many cars, light trucks, and heavy trucks passed over the traffic counter?

Solution: To get a feel for the problem, you might begin by trial and error. For example, you can rule out the possibility of 12 light trucks because 12 light trucks would produce $12 \times 3 = 36$ counts—more than the 35 counts registered. However, 11 light trucks would produce $11 \times 3 = 33$ counts, which, combined with one car, produces the required 35 counts.

Clearly, something more systematic than trial and error is needed. If we let c, t, and T represent the number of cars (two counts), light trucks (three counts), and heavy trucks (four counts), respectively, passing the counter, then the condition that 35 counts are registered becomes

$$2c + 3t + 4T = 35.$$

This single equation in three variables, for which we seek *nonnegative integer solutions*, is an example of a **Diophantine equation**. The theory of Diophantine equations is extensive, but we need little of it to proceed. Rewriting the equation in the form

$$T = \frac{35 - 2c - 3t}{4} = 9 - \frac{1 + 2c + 3t}{4}$$

is useful because it tells us that c and t must be chosen in ways that make $1 + 2c + 3t$ divisible by 4 (recall that T must be an integer). With this observation, we can systematically vary c and t and look for solutions. Table 3.1 shows all possible solutions.

Table 3.1. *The table gives 26 solutions of the traffic counter problem—hardly a unique solution!*

c	t	T	Counts	c	t	T	Counts
0	1	8	35	4	9	0	35
0	5	5	35	5	3	4	35
0	9	2	35	5	7	1	35
1	3	6	35	6	1	5	35
1	7	3	35	6	5	2	35
1	11	0	35	7	3	3	35
2	1	7	35	7	7	0	35
2	5	4	35	8	1	4	35
2	9	1	35	8	5	1	35
3	3	5	35	9	3	2	35
3	7	2	35	10	1	3	35
4	1	6	35	10	5	0	35
4	5	3	35	11	3	1	35

This table has some useful patterns. Notice that the table is generated by increasing the value of c from $c = 0$ to $c = 11$. For each fixed value of c, the corresponding values of t increase by 4, and the corresponding values of T decrease by 3. These patterns provide a valuable check on the solutions; if the pattern is violated, it indicates an error in computation. Finally, this problems has 26 acceptable numerical answers—a nice example of nonuniqueness.   □

Hint 2: More than one method of solution.

Just as we must often admit more than one answer, we should also expect more than one method of solution that leads to an answer. Mathematics and the human mind are far too rich to expect that everyone will follow the same path. As illustrated in the following example, an efficient strategy can save a lot of time and work.

Example 3.2 The second race. Pat and Kris ran a 100-meter race, and Kris beat Pat by 5 meters. They decided to race again, with Kris starting 5 meters behind the start line. Assuming that both runners ran the second race at the same pace as in the first race, who won the second race?

Solution 1: One approach to this problem is analytical. We are not told how fast either Kris or Pat ran, so we can choose some reasonable numbers. For example, we might assume that Kris completed the first race in 20 seconds (s). In that case, her pace was $(100\,\text{m})/(20\,\text{s}) = 5\,\text{m/s}$. Because Pat ran only 95 meters in the same 20 seconds, his pace was $(95\,\text{m})/(20\,\text{s}) = 4.75\,\text{m/s}$. With these rates we can analyze the second race. Kris must run 105 meters in the second race (because she starts 5 meters behind the starting line) to Pat's 100 meters. Their finishing times are

$$\text{Kris: } \frac{105\,\text{m}}{5\,\text{m/s}} = 21\,\text{s} \quad \text{Pat: } \frac{100\,\text{m}}{4.75\,\text{m/s}} = 21.05\,\text{s}.$$

Kris has the faster time and wins the second race.

Solution 2: Although the analytical method works, there is a much more direct and intuitive solution. Here is the **key insight**: Kris runs 100 meters in the same time that Pat runs 95 meters. Therefore, in the second race, Kris will be even with Pat after Pat runs 95 meters. In the remaining 5 meters, Kris' faster speed will allow her to pull away and win. This insight avoids the calculations needed in the first method.   □

Hint 3: Use appropriate tools.

You don't need a supercomputer to check your bill in a restaurant, and you don't need calculus to find the area of a rectangular room. For any given task there is an appropriate level of power that is needed, and it is a matter of style and efficiency to neither underestimate nor overestimate that level. You usually have a choice of tools to use on any problem. Using the tools most suited to the job always makes the task much easier.

Example 3.3 The cars and the canary. Two cars, 120 miles apart, began driving toward each other on a long straight highway. One car traveled 20 miles per hour and the other 40 miles per hour. At the same time a canary, starting on one car, flew back and forth between the two cars as they approached each other. If the canary flew 150 miles per hour and turned around instantly at each car, how far had it flown when the cars collided?

Solution: Because the problem asks for distance, it's tempting to imagine that the canary's journey consists of a sequence of back-and-forth stages. In principle, it is possible to determine the length of each stage and sum the lengths, which amounts to evaluating an infinite (geometric) series. With its use of a limit, this method involves calculus.

The **key insight** is to focus on *time* rather than *distance*. The two cars approached each other with a relative speed of $20+40 = 60$ miles per hour. Together they covered 120 miles in $(120\,\text{mi})/(60\,\text{mi/hr}) = 2\,\text{hr}$, and in 2 hours, the canary flew 300 miles. We see that elementary methods lead to a cleaner and more direct solution.   □

According to legend, the great Hungarian-American mathematician John von Neumann (1900–1955) was given the cars and the canary problem (or some variation of it). As was his style, he solved it instantaneously, and someone observed that he must have used the direct method. He responded that, in fact, he had summed the infinite series in his head!

Hint 4: Similar, simpler problems.

When confronted with a problem that seems daunting, the insight gained from solving a similar, but simpler, problem may help you understand the original problem.

Example 3.4 Coffee and milk. Suppose you have two cups in front of you: one holds coffee and one holds milk. You take a teaspoon of milk from the milk cup and stir it into the coffee cup. Next, you take a teaspoon of the mixture in the coffee cup and put it back into the milk cup. After the two transfers, there will be either: (1) more coffee in the milk cup than milk in the coffee cup; (2) less coffee in the milk cup than milk in the coffee cup; or (3), equal amounts of coffee in the milk cup and milk in the coffee cup. Which of these three possibilities is correct?

Solution: Notice that there is no assumption that the two cups start with the same amount of liquid or that the mixture in the coffee cup is stirred thoroughly after the first transfer. So it is difficult to determine how much coffee and milk are returned to the milk cup after the second transfer. As a simpler analogous problem, you might suppose that the milk cup initially contains 10 white marbles (milk "molecules") and the coffee cup contains 10 black marbles (coffee "molecules"). Also suppose that a teaspoon consists of two marbles. Now you can experiment, as shown in Figure 3.1, with various transfers.

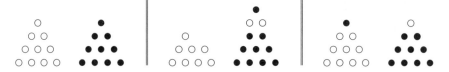

Figure 3.1. *Black and white marbles can be used to simulate the mixing of coffee and milk. The milk cup initially contains 10 white marbles, and the coffee cup contains 10 black marbles (left panel). In the first transfer, two white marbles are moved from the milk cup to the coffee cup (center). In the second transfer, one white marble and one black marble are moved from the coffee cup to the milk cup (right). In the end, the number of black marbles in the milk cup equals the number of white marbles in the coffee cup.*

You can verify that regardless of how the second transfer is done (two white marbles, one white marble and one black marble, or two black marbles), after two transfers, the number of black marbles in the milk cup equals the number of white marbles in the coffee cup. Generalizing, you can also verify that if a teaspoon is defined as three marbles, the outcome is the same. And generalizing further, you can verify that if the cups originally have arbitrary (and different) numbers of marbles at the start, the outcome is the same. While a coffee and milk solution is not modeled precisely by marbles, these preliminary experiments should suggest that of the three options given above, option (3) is correct: The amount of coffee in the milk cup equals the amount of milk in the coffee cup.

This insight leads to a direct solution: Because both cups contain the same amount of liquid before and after the transfers, the milk that is missing from the milk cup has been exactly displaced by coffee, which, in turn, is missing from the coffee cup and has been displaced by the missing milk. Thus, the amount of milk in the coffee cup equals the amount of coffee in the milk cup.   ☐

Seek simplicity and distrust it.
— A.N. Whitehead

Hint 5: Equivalent problems with simpler solutions.

Replacing a problem with a *similar*, simpler problem, as in Hint 4, may not always lead to a complete solution of the original problem. A more powerful strategy is to replace the given problem by an *equivalent*, simpler problem. This is more easily said than done. Finding the equivalent problem or proving the equivalence may be the crux of the problem.

Example 3.5 The ant's journey. Imagine a box-shaped room with an 8-foot ceiling and a rectangular floor that measures 12 feet by 10 feet. An ant is parked on one end wall, one foot from the ceiling and one foot from a side wall. Her goal is to *walk* to a point on the opposite end wall, one foot from the floor and one foot from the opposite side wall. What is the length of the shortest path for the journey?

Solution: The problem is posed in three dimensions, but the ant cannot fly along the shortest straight-line path. The **key insight** is to reduce the problem to two dimensions and allow the ant to walk along the shortest straight-line path. The two-dimensional problem is equivalent to the original problem, and it's much easier to visualize possible paths.

Figure 3.2 shows one unfolding of the room, with the initial point P_0 and the final point P_1 marked. Drawing a line between the points and using the Pythagorean theorem, we see that the length of the journey is

$$|P_0P_1| = \sqrt{20^2 + 8^2} = 21.5 \text{ feet.}$$

With the unfolding trick, the solution is straightforward. But is the answer correct?

Pólya's fourth step advises us to review our work and check for oversights. The twist in this problem is that there are several ways to unfold the room. Figure 3.3 shows several other ways to unfold the room, giving new locations of the start and finish points, which we call P_0' and P_1'. Now there are three new paths with lengths

$$|P_0'P_1| = \sqrt{14^2 + 16^2} = 21.3 \text{ feet,}$$
$$|P_0P_1'| = \sqrt{2^2 + 28^2} = 28.1 \text{ feet,}$$
$$|P_0'P_1'| = \sqrt{6^2 + 22^2} = 22.8 \text{ feet.}$$

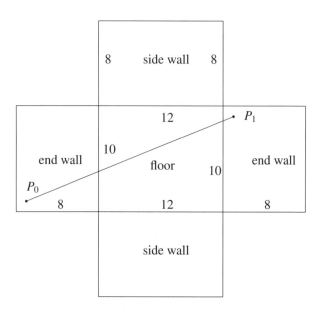

Figure 3.2. *The room in which the ant walks can be unfolded to show one possible straight-line path.*

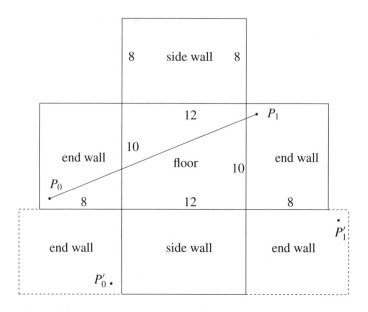

Figure 3.3. *If the end walls of the ant's room are unfolded differently (dashed lines), then three new paths appear. P_0 and P_0' are the initial points in the two unfoldings; P_1 and P_1' are the final points in the two unfoldings. The shortest path is $P_0'P_1$.*

Evidently, the original path from P_0 to P_1 is not the shortest. Instead, the ant should take the path from P_0' to P_1, which means walking across the adjacent side wall, rather than walking down the end wall. Notice that, by the symmetry of the problem, using the ceiling or the other side wall would not produce any new paths. But the moral of the problem is that none of these paths would have been evident without the unfolding of the room.   □

Hint 6: Use units.

Most numbers and variables are measures or counts of some quantity. For this reason, they carry units or dimensions, such as *meters*, *years*, *acre-feet*, or *cubic furlongs*. The use of units is not just a pedantic affair; it is a powerful problem-solving tool that also serves as a necessary (but not sufficient) check of your work.

It's the law: 0.00009 light-years per millennium.

As a simple example, suppose that you plan to take a 450-mile trip in a car that gets 32 miles per gallon of gas, and you expect to pay $2.10 per gallon of gas. How much will the gas for the trip cost? Your intuition may tell you how to combine the three numbers in this problem to get a final answer. But to do the calculation correctly and with conviction, units should be used as follows:

$$\text{Cost of trip} = 400 \text{ miles} \times \frac{1 \text{ gal}}{32 \text{ miles}} \times \frac{\$2.10}{1 \text{ gal}} = \$26.25.$$

There is only one way to combine the numbers in this problem to produce an answer with units of *dollars*, and the units tell you the way. Here is a more substantial example of a very old class of problems.

Example 3.6 Making bagels. The manager of a bagel factory knows that (on average) 3 people can make 450 bagels in 2 hours. Assume that the employees work independently and have the same rate of production.

  (i) How many bagels can 6 people make in 8 hours?
 (ii) How many hours are needed for 2 people to make 800 bagels?
(iii) How many people are needed to make 600 bagels in 2 hours?

Solution: First, let's look at an awkward way of solving the problem by proportions. For part (i), we might reason

- Given: 3 people can make 450 bagels in 2 hours.

- Thus, 6 people can make 900 bagels in 2 hours.

- Thus, 6 people can make 3600 bagels in 8 hours.

Each question can be answered using a variation of this idea. However, units provide a much more unified solution. There are three different variables in the problem: *bagels*, *hours*, and *people*. They can be combined to give a *production rate*, R. If 3 people can make 450 bagels in 2 hours, the production rate is 450 bagels per 3 people per 2 hours, or

$$R = \frac{\frac{450 \text{ bagels}}{3 \text{ people}}}{2 \text{ hours}} = 75 \frac{\text{bagels}}{\text{person} \cdot \text{hour}}.$$

The variables in the problem are related to the production rate by the statement

$$B = \text{Bagels produced} = \text{production rate} \times \text{time} \times \text{people} \quad \text{or} \quad B = RTP.$$

The three questions can now be easily answered . For part (i), we see that

$$B = RTP = 75\frac{\text{bagels}}{\text{person} \cdot \text{hour}} \times 8 \text{ hours} \times 6 \text{ people} = 3600 \text{ bagels}.$$

For part (ii), we have

$$T = \frac{B}{RP} = \frac{800 \text{ bagels}}{75\frac{\text{bagels}}{\text{person·hour}} \times 2 \text{ people}} = 5.33 \text{ hours}.$$

And for part (iii), we have

$$P = \frac{B}{RT} = \frac{600 \text{ bagels}}{75\frac{\text{bagels}}{\text{person·hour}} \times 2 \text{ hours}} = 4 \text{ people}.$$

Notice how the units combine to give dimensionally consistent results. □

Hint 7: Approximations can be useful.

Another useful strategy is to use approximations when appropriate. Most real problems involve approximate numbers to begin with, so a justified approximation often leads to an acceptable answer. In other cases, an approximation will reveal the essential character of a problem, making it easier to reach an exact answer. Approximations can also provide a useful check: If you come up with an exact answer that isn't close to an approximate answer, you ought to be suspicious.

It is the mark of an educated mind to rest satisfied with the degree of precision which the nature of the subject admits and not to seek exactness where only an approximation is possible.
— Aristotle

Example 3.7 The bowed rail. On a particularly hot day, a (hypothetical) railroad rail, originally one mile in length, expands by one foot (see Figure 3.4). Because the ends of the rail are anchored, the rail bows upward along the arc of a circle. How far off the ground is the midpoint of the bowed rail? Look for an approximate solution and then an exact solution.

Solution: Because the added length is short compared to the original length, we can approximate the curved rail with two straight lines, as shown in Figure 3.4. We now have two right triangles. The base of either right triangle is $\frac{1}{2}$ mile $= 2640$ feet long. The hypotenuse of either triangle is $\frac{1}{2}$ mile $+ \frac{1}{2}$ foot $= 2640.5$ feet long. Applying the Pythagorean theorem, we find that the height of the rail off the ground is approximately

$$\sqrt{2640.5^2 - 2640^2} = 51.4 \text{ feet}.$$

Based on our approximation, the center of the rail rises more than 50 feet off the ground! While this result seems surprising, we return to this problem in Chapter 9 and give an exact solution, confirming that the approximation made here is reasonable. □

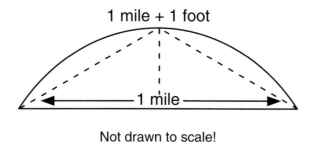

Not drawn to scale!

Figure 3.4. *A one-mile rail expands by one foot and bows upward along the arc of a circle. How high is the midpoint of the rail from the ground? The curved rail can be approximated by straight line segments.*

Hint 8: Change perspective.

Sometimes it helps to assume the perspective of another person or object in the problem, and then ask if the problem is easier to solve. We return to this idea in Chapter 5 when we discuss relative velocity. But here is an example that illustrates the point.

Example 3.8 Coincident clock hands. At what time (exactly) between 1:00 and 2:00 do the hands of a clock coincide?

Solution 1: One of many solutions to the problem begins with the **key insight** that at the moment of coincidence, the two hands have moved for the same length of time starting at 1:00. The position of the hands can be measured in degrees, radians, or revolutions, but it is simplest to measure their position, θ, in minutes of time. Then the initial position of the minute hand (at 1:00) is 0 minutes, and the initial position of the hour hand is 5 minutes. Note that the hour hand moves at $\frac{1}{12}$ the speed of the minute hand. At the moment of coincidence, the minute hand has traveled θ minutes, and the hour hand has traveled $5 + \frac{\theta}{12}$ minutes. Equating these times, we have $\theta = 5 + \frac{\theta}{12}$, or $\theta = \frac{60}{11}$ minutes. The coincidence occurs at $\frac{60}{11}$ minutes past 1:00, or at 1:05:27.

Solution 2: Another solution results if we imagine riding on the hour hand, beginning at 1:00, and looking at the minute hand as it approaches. This time, let's measure the position of the hands in revolutions. Seen from the moving hour hand, the minute hand appears to be approaching at a speed of

$$1 \frac{\text{rev}}{\text{hour}} - \frac{1}{12} \frac{\text{rev}}{\text{hour}} = \frac{11}{12} \frac{\text{rev}}{\text{hour}}.$$

(An analogous situation is riding in a car traveling 30 miles per hour and watching a car approach from behind at 50 miles per hour. The second car appears to approach at $50 - 30 = 20$ miles per hour.) The time required for the minute hand to cover the original separation of $\frac{1}{12}$ of a revolution is

$$\frac{\frac{1}{12} \text{ rev}}{\frac{11}{12} \frac{\text{rev}}{\text{hour}}} = \frac{1}{11} \text{ hour}.$$

Thus, the minute hand catches the hour hand after $\frac{1}{11}$ hours or $\frac{60}{11}$ minutes, as found earlier.   □

Hint 9: Consider special cases.

Often a problem is posed in very general terms, perhaps with one or more parameters. It such situations, it often pays to isolate special cases, which are usually easier to solve. For example, if a problem has parameters, it helps to choose specific values of the parameters. Having solved a few special cases, a general solution may follow more easily.

Example 3.9 Arriving on time. Ralph can reach his destination on time if he averages 60 miles per hour. If at a fraction $0 < p < 1$ of the distance to his destination he realizes that he has averaged only $r < 60$ miles per hour, how fast must he travel to arrive on time?

Solution: Following the advice just given, let's look at the specific case in which $p = \frac{1}{2}$ and $r = 30$ miles per hour. In other words, we will solve the following special case:

> Ralph can reach his destination on time if he averages 60 miles per hour. If at $p = \frac{1}{2}$ of the distance to his destination he realizes that he has averaged only $r = 30$ miles per hour, how fast must he travel to arrive on time?

The familiar distance-rate formula will be useful here. Recall that, when dealing with constant rates or average rates, we know that

$$\text{distance traveled} = \text{speed} \times \text{time elapsed}.$$

Suppose that Ralph needs to travel d miles. At a speed of 60 miles per hour, he can get to his destination on time if he travels no more than $d/60$ hours. We are told that after traveling $d/2$ miles (half the distance), he has averaged only 30 miles per hour, which means that he has traveled for $(d/2)/30 = d/60$ hours. In other words, Ralph has already used all of the time needed to arrive on time; now it's impossible to arrive on time at any finite speed!

This single special case has revealed something important: Depending on the values of r and p in the original problem, the required rate may not be finite. Specifically, if more than halfway to his destination $\left(p \geq \frac{1}{2}\right)$ Ralph realizes that he has averaged 30 miles per hour (or less), he will not be able to arrive on time.

Trying another special case, suppose that at $p = \frac{1}{3}$ of the distance to his destination Ralph realizes that he has averaged only $r = 30$ miles per hour. As before, he can arrive on time only if he travels no more than $d/60$ hours. After traveling $d/3$ miles, he has used $(d/3)/30 = d/90$ hours, which means that he has $d/60 - d/90 = d/180$ hours to travel the remaining $2d/3$ miles. The required speed is $\frac{2d}{3} / \frac{d}{180} = 120$ miles per hour. Legally Ralph may not be able to arrive on time, but mathematically he can! Notice that the answer is independent of d; for clarity, we will retain d in the solution.

Now we can begin to generalize. If at a fraction p of the distance to his destination Ralph realizes that he has averaged only $r = 30$ miles per hour, then he has traveled pd miles at 30 miles per hour. So he has used $pd/30$ hours, which leaves

$$\frac{d}{60} - \frac{pd}{30} = \frac{d}{60}(1 - 2p) \quad \text{hours.}$$

Thus the speed at which he must travel the remaining $(1 - p)d$ miles is

$$s = \frac{(1-p)d}{\frac{d}{60}(1-2p)} = \frac{60(1-p)}{1-2p} \quad \text{miles per hour.}$$

When generalizing, it useful to check that the special cases already considered are recovered. If $p = \frac{1}{2}$, we see that s is undefined, meaning that Ralph cannot arrive on time. If $p = \frac{1}{3}$, then $s = 120$, which is consistent with the previous calculation.

Now we can generalize once more with respect to the parameter r. If Ralph travels pd miles at r miles per hour, then he uses pd/r hours, leaving

$$\frac{d}{60} - \frac{pd}{r} = \frac{d}{60}\left(1 - \frac{60p}{r}\right) \quad \text{hours.}$$

It follows that the speed at which he must travel the remaining $(1 - p)d$ miles is

$$s = \frac{(1-p)d}{\frac{d}{60}\left(1 - \frac{60p}{r}\right)} = \frac{60r(1-p)}{r - 60p} \quad \text{miles per hour.}$$

We now have a general formula that answers Ralph's question for any values of p and r. It is important to check that the three special cases already considered ($p = \frac{1}{2}, r = 30$; $p = \frac{1}{3}, r = 30$; and p arbitrary, $r = 30$) are contained in this general result.

We can also determined under which conditions it is possible and impossible for Ralph to arrive on time. Notice that the numerator in the general speed formula is positive for $0 < p < 1$. Because s must be positive, we obtain physically meaningful solutions, provided that the denominator is positive, which means that $r > 60p$ or $p < r/60$. Thus Ralph can arrive on time, provided that he notices his leisurely speed of r miles per hour at no more than a fraction $p = r/60$ of the distance to his destination. The regions of the (p, r) parameter plane in which Ralph can and cannot arrive on time are shown in Figure 3.5. For example, if at a fraction $p = 0.4$ of the way to his destination Ralph realizes he has been averaging $r = 25$ miles per hour, then he can arrive on time. However, if at a fraction $p = 0.4$ of the way to his destination Ralph realizes he has been averaging $r = 10$ miles per hour, then he cannot arrive on time.   □

Hint 10: Get the picture.

We have already presented several examples in which a picture is helpful, if not indispensable. If you are a visual thinker, then this hint already makes sense to you. However, nonvisual thinkers may also find that drawing pictures helps organize information and suggests solution paths.

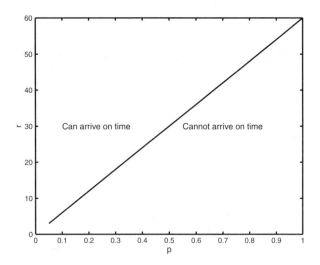

Figure 3.5. *The graph shows the values of p and r for which Ralph can and cannot arrive on time. If $r > 60p$, then Ralph can arrive on time.*

Example 3.10 Double round trip. Jim left Abletown (A) at noon on a bicycle and rode to Bakersville (B) and back, a round-trip distance of 52 miles. Sometime after noon, Bob left Bakersville and drove to Abletown and back on the same road, completing the round trip at 3:20 p.m. Jim and Bob passed each other 7.5 miles from Bakersville on Jim's outbound trip, and again 5.5 miles from Bakersville on Jim's return trip. Both men traveled at constant (but different) speeds throughout their journeys. When did Jim return to Abletown? [17]

Solution: This story is filled with detailed information. A picture helps tremendously in organizing the information. Figure 3.6 shows the *physical* map of the double round trip with all the given information.

It is now possible to make a few preliminary observations. Note that between the two passing points, Jim traveled 7.5 miles (to B) plus 5.5 miles (from B), or 13 miles. At the same time, Bob traveled $(26 - 7.5)$ miles (to A) plus $(26 - 5.5)$ miles

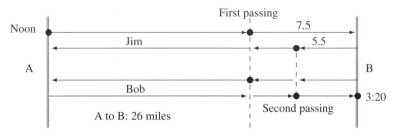

Figure 3.6. *In the double-round-trip problem, Jim left A at noon and returned to A at a time to be determined. Bob left B at an unspecified time and returned to B at 3:20 p.m.*

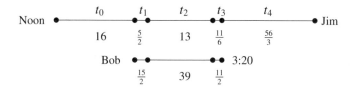

Figure 3.7. *The time map of the double round trip shows that Jim's trip takes place in five stages of length $t_0, \ldots, t_4$ hours. Bob's trip takes place in three stages of length t_1, t_2, t_3 hours. The numbers below each time line give the distances traveled in each stage.*

(from A) for a total of 39 miles. Thus, in the same period of time, Bob traveled three times further than Jim. Letting v_J and v_B denote the speeds of Jim and Bob, respectively, we have $v_B = 3v_J$. Thus we also see that in the time that Bob traveled the 7.5 miles from B to the first passing point, Jim traveled $\frac{7.5}{3} = 2.5$ miles. Similarly, in the time that Bob traveled the 5.5 miles from the second passing point to B, Jim traveled $\frac{5.5}{3} = \frac{11}{6}$ miles. (This also implies that Jim traveled 16 miles at the beginning of his journey before Bob started and $\frac{56}{3}$ miles at the end of his journey after Bob finished.) These conclusions are summarized on the time lines shown in Figure 3.7.

Notice that the time elapsed between Jim's departure from A and Bob's return to B is 3 hours and 20 minutes, or $\frac{10}{3}$ hours. In this time, Jim traveled $\left(26 + \frac{11}{2} + \frac{11}{6}\right) = \frac{100}{3}$ miles. This means that his speed was $v_J = \frac{100}{3} / \frac{10}{3} = 10$ miles per hour. Thus, the time required for Jim's entire trip was (52 miles)/(10 miles per hour) = 5.2 hours or 5 hours and 12 minutes. Jim returned to Abletown at 5:12 in the afternoon.   □

Hint 11: Try alternative thought patterns.

Try to avoid rigid patterns of thought that tend to suggest the same ideas and methods. Rather, try to approach every problem with an openness and freshness that encourages innovative ideas. In its most wondrous form, this approach is typified by what Martin Gardner calls *AHA!* solutions. Such solutions involve a penetrating and unexpected revelation that reduces the problem to its essential parts.

Example 3.11 The monk and the mountain. A monk set out from a monastery in the valley at dawn. He walked all day up a winding path, stopping for lunch and taking a nap along the way. At dusk he arrived at a temple on the mountaintop. The next day the monk made the return walk to the valley, leaving the temple at dawn, walking the same path for the entire day, and arriving at the monastery in the evening. Must there be one point along the path that the monk occupied at the same time of day on both the ascent and descent? [18]

Solution: There is a temptation for mathematically inclined people to devise functions for the monk's journeys up and down the mountain. Then an application of the intermediate value theorem eventually confirms that, indeed, there is point along the path that the monk occupied at the same time of day on both journeys. Far simpler is to forget that the journeys occurred at two different times; instead imagine the monk and his twin making the ascent and descent simultaneously. There can be no doubt that the two monks must pass each other at the same point along the trail—at the same time of day.   □

Hint 12: Don't spin your wheels.

Often you get so tangled up in a problem that you couldn't see a solution if it were right before your eyes. It's easier said than done, but when your wheels are spinning, let up on the gas! Often the best strategy in problem solving is to put a problem aside for a few hours or days. You will be amazed at what you may see (and what you overlooked) when you return to it.

In the following problems, work with Pólya's method, use the hints of this chapter, and try to observe yourself in the act of problem solving. Write your solutions clearly, and comment on the problem-solving process.

3.1   Exercises

3.1. ◇ Ticket confusion. Alex is selling tickets for the spring orchestra concert. There are three ticket categories: student tickets for $10, regular tickets for $20, and senior tickets for $15. At the end of the evening, Alex has collected $125, but has no record of how many tickets were sold in each category. How many student, regular, and senior tickets could have been sold?

3.2. Diophantine barnyard. A farmer visits his barnyard to find that the cows and the roosters are socializing. He counts a total of 88 legs in the barnyard.

  (i) Find a general formula that gives all possible sets of cows and roosters that the farmer could have seen.

  (ii) Realizing that he cannot determine the exact number of cows and roosters, the farmer also counts 32 heads in the barnyard. How many cows and roosters are in the barnyard?

3.3. ◇ Buy and sell. A woman bought a horse for $500 and then sold it for $600. She bought it back for $700 and then sold it again for $800. How much did she gain or lose on these transactions?

3.4. Another ant's journey. Suppose that the ant in the $12' \times 10' \times 8'$ room of Example 3.5 starts at a point on an end wall 2 feet from the ceiling and 2 feet from a side wall. She wants to *walk* to a point on the opposite end wall 2 feet from the floor and 2 feet from the opposite side wall. What is the shortest path?

3.5. ◇ Coiled wire. Suppose that eight turns of a wire are wrapped uniformly around a pipe with a length of 20 centimeters and a circumference of 6 centimeters. What is the length of the wire?

3.6. Handshakes. Every person on earth has shaken a certain number of hands. Prove that an even number of people have shaken an odd number of hands.

3.7. ◇ The bookworm.

  (i) Two identical volumes of Shakespeare rest on a bookshelf in the natural order (volume 1 on the left and volume 2 on the right as you face the bookshelf). The volumes are one inch thick, and the covers are each one-quarter of an inch thick. A bookworm begins on the first page of the first volume and burrows in a straight line to the last page of the second volume. How far does the bookworm travel?

  (ii) The complete works of Shakespeare in n identical volumes rest on a bookshelf. The volumes are one inch thick and the covers are each one-quarter of an inch thick. A bookworm begins on the first page of the first volume and burrows in a straight line to the last page of the nth volume. How far does the bookworm travel?

3.8. High stakes. Two friends, A and B, play a game several times. The stake is one penny per game; this means that the winner of a game gains one penny and the loser loses one penny. At the end of the day, A's net gain is three pennies, and B has won three games. How many games did they play? [14]

3.9. ◇ Extra circumference. Imagine a string stretched snugly about the equator of a precisely spherical Earth. Now imagine increasing the length of this string by 2π feet and pulling the new string away from the Earth the same distance everywhere along its length. What is the (constant) distance between the new string and the Earth? Assume that the radius of the Earth is 4000 miles.

3.10. Shrink-wrapped Earth. Now suppose that a precisely spherical Earth is tightly wrapped in a sheet of plastic. How many square feet of plastic must be added to the original sheet so that the new sheet is everywhere one foot off the surface of the Earth? Assume that the radius of the Earth is 4000 miles. [20]

3.11. ◇ Suspicious survey. A survey shows that of 100 nurses, 75 play at least soccer, 95 play at least softball, and 50 play both soccer and softball. Is this possible?

3.12. Coins in pockets. Bob has 10 pockets and 44 silver dollars. Can he put his dollars in his pockets so that no two pockets have the same number of dollars? [29]

3.13. ◇ Auto auction. At a classic car auction, 30 buyers were present. Ten buyers bought fewer than 6 cars. Eight buyers bought more than 7 cars. Five buyers bought more than 8 cars. One buyer bought more than 9 cars. How many buyers bought 6, 7, 8, or 9 cars? [39]

3.14. Senior citizens. The annual dues for a book club are $23 per person, with a special rate of $17 for senior citizens. If a total of $1500 is collected in dues for the year, what is the smallest number of senior citizens that could belong to the club? [1]

3.15. ◇ Eavesdropping. You overhear the following conversation:

Paul: How old are your three children?

Paula: The product of their ages is 36, and the sum of their ages is the same as today's date.

Paul: That is not enough information.

Paula: The oldest child also has red hair.

If you were Paul, could you determine the ages of Paula's children? Explain.

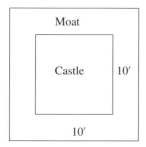

3.16. Crossing the moat. A castle is surrounded by a deep moat that is 10 feet wide (see figure above). A knight on a rescue mission must cross the moat with only two $9\frac{1}{2}$-foot planks. Without using glue, nails, or supernatural means, how does he do it?

3.17. ◇ Surface area. If one pint of paint is needed to paint a statue six feet tall, then how many pints are needed to paint (with the same thickness of paint) 540 statues similar to the original, but scaled down so they are only one foot tall? [24]

3.18. Ants in pursuit. Four ants are at the corners of a ten-inch square. At the same moment, each ant begins crawling toward its neighbor to the left at the same constant rate. How far do the ants crawl before they all meet? Try a change of perspective. [10]

3.19. ◇ An old chestnut. A ship is twice as old as the ship's boiler was when the ship was as old as the boiler is now. What is the ratio of the ship's age to the boiler's age?

3.20. Counting handshakes. Mo attended a party at which there were 23 people and observed that at least two people shook hands with the same number of other people. Of course, no one shook hands with the same person twice and no one shook his/her own hand. Mo could not have counted every handshake that took place. Could she be correct?

3.21. ◇ Commuting. A woman can commute to work by train or bus. If she takes the train in the morning, she takes the bus in the afternoon. If she takes the train in the afternoon, it means she took the bus in the morning. During a total of n working days, the woman took the bus in the morning 8 times, took the bus in the afternoon 15 times, and took the train (either way) 9 times. What is n? [24]

3.22. Coincident clock hands. (i) At what time between 2:00 and 3:00 is the minute hand directly over the hour hand? (ii) At what time between n:00 and $(n+1)$:00 is the minute hand directly over the hour hand?

3.23. ◇ Single elimination. A ping-pong tournament is organized as follows: The name of each contestant is put in a hat, and then names are drawn out in pairs. The people in each pair play each other, with the loser retiring from the tournament. The names of the winners are once again placed in a hat, and names are drawn in pairs to determine the games of the second round. This process continues until only the winner remains. If at any stage there is an odd number of names in the hat, the undrawn name remains in the hat until the next round. If N people enter the tournament, how many games are needed to determine a winner?

3.24. Clock hands. Henry started a trip between 8:00 AM and 9:00 AM, when the hands of his watch were together. He arrived at the destination between 2:00 PM and 3:00 PM, when the hands of his watch were exactly opposite each other. How long did he travel? [30]

3.25. ◇ Coin collectors. At a meeting of a coin collectors' club, each member present had 50 U.S. coins (pennies, nickels, dimes, quarters, and half-dollars). Each person's coins added up to $1, and no two members had exactly the same combination of coins. What is the largest number of people who could have attended the meeting? [1]

3.26. Catch and release. Suppose you want to estimate the number of fish in a large lake. Imagine catching 50 fish, marking them in some harmless way, and then releasing them back into the lake. Several days later, after the marked fish have dispersed throughout the lake, you catch another 50 fish and discover that 2 of these fish are tagged. What is the best estimate for the number of fish in the lake?

3.27. ◇ Security system. An airport security system consists of two stages. In the first stage, 50% of travelers pass without inspection. In the second stage, 75% of travelers pass without inspection. What fraction of travelers are inspected zero times, once, and twice?

3.28. The double window. Imagine a double window with the upper darker window fixed and the lower lighter window capable of sliding from fully closed (no overlap with the upper window) to fully open (complete overlap with the upper window). When the lower window is half open, the total amount of light passing through the combined window is 1% less than when the lower window is fully open. When the lower window is half open, the combined window stops 5% more light than when the lower window is fully open. If the two windows have the same area, (i) what fraction of the light falling on the lower window is stopped by the window? (ii) what fraction of the light falling on the upper window is stopped by the window? [17]

3.29. ◇ Six flasks. Suppose that you have six flasks with capacities of 16, 18, 22, 23, 24, and 34 ounces. You fill some of the flasks with water and all but one of the remaining flasks with wine, using twice as much wine as water. Which flask was not filled, and which flasks were used for water and wine? [17]

3.30. Messy desktop. Fifteen sheets of paper of various sizes and shapes lie scattered on a desktop, covering it completely. The sheets may overlap each other or hang over the edge of the desktop. Is it always true that five of the sheets may be removed so that the remaining ten sheets cover at least two-thirds of the desktop? Explain. [19]

3.31. ◇ Cube in cone. A cube is inscribed in a right circular cone with a radius of 1 and a height of 3. What is the volume of the cube? Use a good picture, but no calculus.

3.32. Cube in sphere in cone in cylinder. A cube with side length r is inscribed in a sphere, which is inscribed in a right circular cone, which is inscribed in a right circular cylinder. The side length (slant height) of the cone is equal to its diameter. What is the surface area of the cylinder? Use a good picture, but no calculus. [20]

3.33. ◇ Hole in a sphere. A cylindrical hole 10 inches long is drilled symmetrically through the center of a sphere. How much material is left in the sphere? Use a good picture, but no calculus.

3.34. A safe move. To move a large heavy safe, two cylindrical bars, 7 inches in diameter, are used as rollers. Assuming no slipping, how far forward does the safe move when the bars have made one complete revolution? [14]

3.35. ◇ Party handshakes. Mr. and Mrs. Schmidt had a party and invited four other couples. Nobody at the party shook hands with the same person twice, and nobody shook the hand of his/her spouse or his/her self. At the end of the evening, Mr. Schmidt polled each person to determine how many hands each person shook. Each person gave a different number of handshakes. How many hands did Mrs. Schmidt shake? [13]

3.36. A surprising result. The 100 numbers in the set $\left\{1, \frac{1}{2}, \frac{1}{3}, \ldots, \frac{1}{100}\right\}$ are written on a blackboard. One may erase any two numbers, a and b, and replace them by the single number $a + b + ab$. After 99 such operations, only one number remains. What is this number? [provided by Stan Wagon, Macalester College]

3.2 Hints and Answers

3.1. HINT: Find the relationship that describes the number of tickets sold. Solve for the number of student tickets sold. ANSWER: There are 16 different solutions; see the Solution section for the list.

3.2. HINT: Find the relationship between the number of legs of each animal type. ANSWER: (i) Solutions satisfy $r = 44 - 2c$ for $c = 0, 1, 2, \ldots, 22$. (ii) There are 12 cows and 20 roosters.

3.3. HINT: Record the transactions in the woman's checkbook. ANSWER: She gained $200 on the transactions.

3.4. HINT: Draw a good picture and consider all possible paths. ANSWER: The shortest distance is 20.9 feet.

3.5. HINT: Transform the problem into a two-dimensional problem.

ANSWER: The length of the wire is 52 centimeters.

3.6. HINT: Draw pictures for small numbers of people and look for a pattern.

3.7. HINT: Draw good pictures and carefully identify the start and end of the worm's journey.
ANSWER: (i) The path through two volumes has a length of $\frac{1}{2}$ inch. (ii) The path through n volumes has a length of $\frac{3}{2}n - \frac{5}{2}$ inches.

3.8. HINT: Let n and $m = 3$ be the number of games won by A and B, respectively.
ANSWER: The friends played nine games.

3.9. HINT: You do not need to compute the circumference of the Earth.
ANSWER: The new string is one foot from the Earth everywhere.

3.10. HINT: The surface area of a sphere is $S = 4\pi r^2$.
ANSWER: With 19 additional square miles of plastic, the sheet is everywhere one foot away from the surface of the Earth.

3.11. HINT: How many nurses play soccer only and softball only?
ANSWER: The survey is flawed.

3.12. HINT: What is the least number of dollars that can be put in ten pockets so that no two pockets have the same number of dollars?
ANSWER: It cannot be done.

3.13. HINT: Identify the unnecessary information.
ANSWER: Nineteen buyers bought 6, 7, 8, or 9 cars.

3.14. HINT: Find the relationship between the number of regular and senior members.
ANSWER: The least number of senior citizens is 3.

3.15. HINT: Each statement has essential information.
ANSWER: The children have ages 2, 2, and 9.

3.16. HINT: Consider the corners of the moat.
ANSWER: Place the first plank across a corner of the moat. Place the second plank from the first plank to the castle.

3.17. HINT: Paint covers surface area, which scales as the square of the linear dimensions.
ANSWER: The miniature statues require 15 pints of paint.

3.18. HINT: Look at the problem from the perspective of one of the ants.
ANSWER: Each ant crawls ten inches.

3.19. HINT: Read carefully and find two equations for the current age of the ship and boiler.
ANSWER: The ratio of the ship's age to the boiler's age is $\frac{4}{3}$.

3.20. HINT: Assign a different number of handshakes to each person.
ANSWER: Yes, she was correct, independent of the number of people at the party.

3.21. HINT: There are four categories of round trips: train/bus and morning/afternoon.
ANSWER: There were 16 working days.

3.22. HINT: There are several solutions. One measures the position of the hands in minutes (60 minutes is one revolution). Another takes the perspective of someone sitting on the hour hand.
ANSWER: (i) The hands coincide at 2:10:54. (ii) The hands coincide at $\frac{60n}{11}$ minutes after n:00.

3.23. HINT: How many contestants are eliminated?
ANSWER: The tournament consists of $N - 1$ games.

3.24. HINT: Determine the position of the hands at the start and end of the trip.
ANSWER: The trip lasted six hours.

3.25. HINT: Write the linear relationships for the number of pennies, nickels, dimes, quarters, and half-dollars.
ANSWER: At most two people could have attended the meeting.

3.26. HINT: Assume that the proportion of marked fish in the second batch of 50 fish equals the proportion of marked fish in the entire lake.
ANSWER: The best estimate for the number of fish in the lake is 1250.

3.27. HINT: Assume that N people pass through the security system, and compute the number of people stopped at each stage.
ANSWER: One-eighth of the travelers are inspected twice, $\frac{1}{2}$ of the travelers are inspected once, and $\frac{3}{8}$ of the travelers are not inspected.

3.28. HINT: Let x be the fraction of light that the upper window allows to pass. Let y be the fraction of light that the lower window allows to pass, where $0 < x < y < 1$.
ANSWER: (i) The upper (darker) window stops one-fifth of the light. (ii) The lower (lighter) window stops one-sixth of the light.

3.29. HINT: Show that, after neglecting one of the flasks, the sum of the capacities of the remaining flasks is divisible by 3.
ANSWER: The 23-ounce flask was not filled, the 16- and 22-ounce flasks were filled with water, and the 18-, 24-, and 34-ounce flasks were filled with wine.

3.30. HINT: Imagine taking one piece of paper the size of the desktop and cutting it arbitrarily into fifteen pieces. Or imagine painting the desk before putting the sheets of paper on the desk.
ANSWER: It is always true: Remove the five sheets of paper with the least direct contact with the desktop.

3.31. HINT: Slice the cone along its vertical axis through the diagonal of the base of the cube. Then draw good pictures.

ANSWER: The volume of the cube is $\left(\frac{6}{2+3\sqrt{2}}\right)^3 \doteq 0.888$.

3.32. HINT: Draw a good cross section of the four solids and identify an equilateral triangle.

ANSWER: The surface area of the cylinder is

$$\frac{9\pi r^2}{2}(1+\sqrt{3}).$$

3.33. HINT: Either use the fact that the radii of the sphere and hole are not given, or use geometry to compute the volume directly.

ANSWER: The amount of material removed is $\frac{500\pi}{3}$ cubic inches.

3.34. HINT: Determine the distance moved by the rollers without the safe, and the distance moved by the safe due only to the revolving of the rollers.

ANSWER: The safe moves 14π inches.

3.35. HINT: What is the maximum number of hands that any person shook? How many hands were shaken by the spouse of the person who shook eight hands?

ANSWER: Mrs. Schmidt shook four hands.

3.36. HINT: Begin with $\left\{1, \frac{1}{2}, \frac{1}{3}\right\}$, and then consider two numbers of the form $\left\{n, \frac{1}{n+1}\right\}$.

ANSWER: The result is 100, independent of the order in which numbers are eliminated.

Chapter 4

How Do You Do It?

*That is what learning is. You suddenly under-
stand something you've understood all your life,
but in a new way.*

— Doris Lessing

The problems of this chapter have a slightly different flavor than those in other chapters. Whereas many problems ask *how many?* or *how far?* or *how much?*, the problems of this chapter simply ask *how?*. These are procedural or process questions, often with extra emphasis on how to do something in the most efficient way possible. The examples and exercises that follow may seem to involve little mathematics, but they do require considerable ingenuity. Furthermore, they point to several important areas of mathematics, such as optimization, operations research, and game theory, whose goal is to design optimal or winning procedures. We will proceed by way of examples that demonstrate a few typical strategies—all within the framework of Pólya's method.

It's better to know some of the questions than all of the answers.
— James Thurber

Example 4.1 Counterfeit golf balls. Each of ten large barrels is filled with golf balls that all look alike. The balls in nine of the barrels weigh one ounce, and the balls in one of the barrels weigh two ounces. You have a scale that measures absolute weight in ounces. With only *one* weighing on this scale, how can you determine which barrel contains the heavy golf balls?

Solution: The **key insight** is that the one allowed weighing must somehow include information from all ten barrels. Furthermore, each barrel must be represented in a unique way in the weighing in order to identify the barrel with the heavy balls. There are several ways to choose the balls for the weighing. Perhaps the easiest is to take one ball from the first barrel, two balls from the second barrel, and so forth, up to ten balls from the tenth barrel. In this way, there will be $1 + 2 + \cdots + 10 = 55$ balls on the scale, which would weigh 55 ounces *if* all the balls weighed one ounce. However, if the 55 balls weigh n ounces over 55 ounces, it means that there are n heavy balls on the scale, which means that barrel n contains the heavy balls. □

Lewis Carroll was the pseudonym of the mathematician and writer Charles Lutwidge Dodgson (1832–1898). He is best known for his two children's books *Alice's Adventures in Wonderland* (1865) and *Through the Looking Glass* (1872). He was a master of puzzles and problems of logic.

Example 4.2 A Lewis Carroll favorite. A queen, her son, and her daughter are captive in the tower of a castle. Outside their window is a rope running over a pulley, with baskets of equal weight attached to the ends of the rope. One basket is outside the window and is empty. The other basket is on the ground and contains a 30-kilogram rock. There is enough friction in the pulley so that one basket can be lowered to the ground using the other basket as a counterbalance, provided that the weight difference between the two baskets does not exceed 6 kilograms. If the weight difference is greater than 6 kilograms, then the heavy basket will crash to the ground. The queen weighs 78 kilograms, the daughter weighs 42 kilograms, and the son weighs 36 kilograms. Each basket can hold either two people or one person and the rock. Describe how the queen and her children can escape to the ground in the fewest number of steps. [10]

Table 4.1. *Tower escape. Q = queen, D = daughter, S = son, R = rock.*

Step	0	1	2	3	4	5	6	7	8	9
Tower	QDS	QDR	QSR	QS	DSR	DS	DR	SR	S	R
Ground	R	S	D	DR	Q	QR	QS	QD	QDR	QDS

Solution: The escape will be accomplished by successively lowering people to the ground using various combinations of people and the rock as a counterbalance. The key constraint is the six-kilogram weight difference between the two baskets. The following combinations can be used to lower a basket: the son and the rock; the son and the daughter; and the rock and the daughter opposing the queen. From here, a bit of trial and error is needed. A good system of recording the trips is also useful. The diagram in Table 4.1 illustrates the various trips that are needed. Note that in three steps, the rock is dropped to the ground. We see that ten steps are needed, and after the tenth step the rock is poised to fall on the escaped party!   □

Example 4.3 Measuring. Measuring or decanting problems are ancient and legion. In these problems, it is generally understood that the given containers have no markings for measuring fractions of the container capacity. This means that you can fill a container to capacity, empty it entirely into another container, or discard its contents. Here is just one of many such problems. Given a 3-gallon jug and a 5-gallon jug with no markings, how can you obtain 1 gallon of water from a well?

Solution: Often a bit of trial and error is enough to solve measuring problems. However, such an approach may not indicate whether you have found the only method or the best method. One sequence of five transfers is shown in Table 4.2. We use the notation (a, b) to indicate that the 3-gallon jug contains a gallons and the 5-gallon jug contains b gallons.

There is a nice algebraic way to analyze measuring problems like the one just presented. It also answers the question of when such measuring tasks are possible. In general, suppose that we have two jugs with capacities of J and K gallons, and suppose that we are asked to finish with G gallons in one of the jugs; in the above example, we have $J = 3$, $K = 5$, and $G = 1$.

Table 4.2. *Measuring 1 gallon using a 3-gallon jug and a 5-gallon jug.*

Jug status	Move
(0,0)	Both jugs empty
(3,0)	Fill 3-gallon jug from well
(0,3)	Empty 3-gallon jug into 5-gallon jug
(3,3)	Fill 3-gallon jug from well
(1,5)	Fill 5-gallon jug from 3-gallon jug
(1,0)	Empty 5-gallon jug into well

There are three basic moves in problems of this sort: fill an empty jug from the well, empty a full jug into the well, or empty a full jug into another jug. We will use two counters, p and q, for the J-gallon jug and the K-gallon jug, respectively; they begin with a value of zero. Each time a jug is filled from the well, we increase the corresponding counter by 1, and each time we empty a jug into the well, we decrease the corresponding counter by 1. At the end of the process, p and q could be negative or positive. And at the end of the process, after a net p fillings of the J-gallon jug and a net q fillings of the K-gallon jug, we must have

$$pJ + qK = G,$$

where J, K, and G are given. The question is whether there are integers p and q that satisfy this equation (another Diophantine equation). For the example above, the relevant equation is

$$3p + 5q = 1.$$

Some trial and error reveals that one solution is $p = 2$ and $q = -1$, which corresponds to the procedure described in Table 4.2, in which the 3-gallon jug was filled from the well twice, and the 5-gallon jug was emptied into the well once. Notice that filling one jug from another jug does not change the values of p and q. Also the values of p and q do not tell us the order in which the jugs are filled and emptied. Another solution is $p = -3$ and $q = 2$. You should verify that this solution corresponds to a nine-step procedure.

Some measuring tasks are not possible. For example, if asked to measure 1 gallon from a well with a 2-gallon jug and 4-gallon jug, you should not spend too much time trying! Inspection of the corresponding equation

$$2p + 4q = 1$$

shows that there are no integer solutions, because the left side is always even.   □

Everything that can be invented has been invented.
— Charles H. Duell, Commissioner, U.S. Office of Patents, 1899

Example 4.4 Coin weighing. Coin weighing problems are numerous (see the Exercises), and this one captures some of the strategy required to solve them. You are given four coins that appear to be identical and told that one of them is a counterfeit; it may be heavy or it may be light. You are also given a fifth coin and told that it is a good coin. Given a balance scale, how can you find the counterfeit coin and determine whether it is heavy or light in two weighings? To be clear, each weighing determines which pan has the heavier sample, but it does not provide absolute weights.

Solution: The strategy in all coin weighing problems is to reduce as much as possible the total number of cases that need to be examined with each weighing. Even with this insight, the solution inevitably requires a bit of trial and error.

Imagine labeling the possible counterfeit coins C_1, C_2, C_3, C_4, and the good coin G. Suppose you put C_1 and C_2 on one pan of the balance and C_3 and C_4 on the other pan. Because one of the coins is counterfeit, the pans will not balance. But what have you learned? Without knowing whether the counterfeit is heavy or light, it is impossible to tell whether a light counterfeit made its pan rise or a heavy counterfeit made its pan drop. This false start says that the good coin must be involved in the first weighing.

Suppose you weigh the good coin against one of the possible counterfeits. In the worst case, four weighings are needed to find the counterfeit. This false start says that at least two coins must be on each pan for the first weighing.

Suppose you put C_1 and G on the right pan and C_2 and C_3 on the left pan. There are three outcomes:

- If the pans balance, then all four coins are good and the counterfeit is C_4. With one more weighing of C_4 against G, you can determine whether C_4 is heavy or light, and you are done.

- If the right pan drops (left pan rises), you can conclude that *either* C_1 is heavy *or* that C_2 or C_3 is light.

- If the right pan rises (left pan drops), you can conclude that *either* C_1 is light *or* that C_2 or C_3 is heavy.

Consider Case 2 in which *either* C_1 is heavy *or* C_2 or C_3 is light. If you now weigh C_2 against C_3 and the pans balance, then C_1 is a (heavy) counterfeit. If the pans do not balance, the pan that rises has a (light) counterfeit coin. Case 3 is resolved in a similar way. Thus two weighings suffice to find the counterfeit and determine whether it is heavy or light.   □

4.1 Exercises

4.1. Crossing problems.

(i) (From the eighth century). A man arrived at the bank of a river with a wolf, a goat, and a head of cabbage. His boat holds only himself and one of his possessions. Furthermore, the goat and the cabbage cannot be left alone, and the wolf and the goat cannot be left alone. What is the minimum number of trips needed for the man to cross the river with his three possessions? Show how it is done.

(ii) ◇ (Tartaglia's Bride Problem). This time three beautiful brides and their jealous husbands come to a river. The ferry boat holds only two people. Furthermore, no woman may be alone with a man unless her husband is present. How can all three couples cross the river in the minimum number of trips?

4.2. Measuring problems. In all of these problems, the containers have no markings, so fractions of the full volume cannot be measured.

(i) ◇ How do you measure exactly 2 gallons of water from a well using a 4-gallon jug and a 7-gallon jug?

(ii) How do you measure exactly 2 gallons of water from a well using a 7-gallon jug and a 13-gallon jug?

(iii) ◇ How do you measure exactly 2 gallons of water from a well using a 7-liter jug and an 11-liter jug?

(iv) Two friends have an 8-quart jug of wine that they want to share. They also have an empty 3-quart jug and an empty 5-quart jug. How can the wine be divided so there are 4 quarts in each of the two larger jugs?

(v) ◇ How do you measure 9 minutes with a 7-minute hourglass and a 4-minute hourglass?

(vi) How do you measure 30 minutes with a 9-minute and a 13-minute hourglass? At least one hourglass must be turned for the first time at the beginning of the 30-minute interval; that is, no start-up period is allowed.

4.3. ◇ Measuring with cylinders.

(i) How can a clear *cylindrical* 16-ounce glass be used to measure exactly 8 ounces of water (with no other measuring devices)?

(ii) Three moonshiners wanted to share the 24 quarts of whiskey in a large barrel. The only available measuring devices were three cylindrical containers holding 10, 11, and 13 quarts. The containers were marked as to their full content, but had no markings for individual quarts. How did the moonshiners measure out three containers with 8 quarts in each? [25]

4.4. Coin weighing problems. In the following problems, assume that a balance scale is used and that all coins in question have identical appearance.

(i) ◇ How do you find the light counterfeit coin among eight coins in two weighings?

(ii) How do you find the heavy counterfeit coin among twelve coins in three weighings?

(iii) ◇ How do you find the light counterfeit coin among 80 coins in four weighings? [20]

(iv) Is it possible to find the light counterfeit coin among thirty coins in three weighings? Explain.

(v) ◇ How do you identify two counterfeit coins, one heavy and one light, among five coins in three weighings?

(vi) How do you find the counterfeit coin (heavy or light) among nine coins in three weighings and determine whether it is heavy or light?

(vii) ◇ How do you find the counterfeit coin (heavy or light) among 12 coins in three weighings and determine whether it is heavy or light?

(viii) How do you determine the order of weights of five coins with different weights in seven weighings? [1]

(ix) ◇ Assume that you have 24 coins that are identical in appearance, n made of steel and $24-n$ made of brass. The steel coins are heavier than the brass coins. How many weighings are needed to determine n? [20]

(x) How do you identify the individual weights of five coins that weigh 10, 20, 30 , 40, and 50 grams in five weighings? To be clear, each weighing determines which pan has the heavier sample, but it does not provide absolute weights. [1]

4.5. ◇ Weighing with weights. Six coins are identical in appearance, but one is counterfeit, either heavy or light. You have an accurate scale that measures absolute weight in ounces. Find a procedure for determining the weights of all six coins in only three weighings. [19]

4.6. Measuring bricks.

(i) Suppose you have one rectangular brick, a pencil, and a ruler. How do you measure the length of the longest diagonal of the brick without using a mathematical formula?

(ii) Suppose you have three identical rectangular bricks and a ruler. In a way different than in part (i), how do you measure the length of the longest diagonal of the brick without using a mathematical formula? [40]

4.7. ◇ The three switches. Suppose you are in a room that has three light switches, each connected to exactly one lamp in an upstairs room. How can you determine which switch operates each lamp with only one trip to the upstairs room? Assume that the up position of the switch is on.

4.8. Blacksmiths unite. One blacksmith needs five minutes to put on one horseshoe. Can eight blacksmiths shoe ten horses (40 horseshoes) in less than 30 minutes? (A horse must stand on at least three legs.) [40]

4.9. ◇ Moving coins. How can you move only three of the coins in the triangle below to make the triangle point downwards?

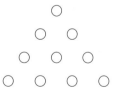

4.10. Weighing golf balls. Each of ten large barrels is filled with golf balls that all look alike. The balls in nine of the barrels weigh one ounce and the balls in one of the barrels weigh two ounces. You have a scale that measures weight to the nearest *pound*. With only one weighing on this scale, how can you determine which barrel contains the heavy golf balls? (Compare to Example 4.1, and note that one pound is sixteen ounces.)

4.11. ◇ **Sanitary surgeons**. How can three surgeons operate successively on a highly infected patient using only two pairs of sterilized gloves? A glove that has touched a surgeon cannot touch the patient or another surgeon, and a glove that has touched the patient cannot touch a surgeon. [10]

4.12. Optimal grilling time. A small grill can hold two hamburgers at a time. If it takes five minutes to cook one side of a hamburger, what is the shortest time need to grill both sides of three hamburgers?

4.13. ◇ **Outsmarting the bus driver**. A boy carrying a brand new five-foot fishing pole is told he cannot ride the bus with an object that has a dimension of more than four feet. He goes back into the store where he bought the pole and explains his dilemma. The owner of the store gives him something that solves his problem. When the next bus comes, the boy gets a ride with his fishing pole in one piece, with no questions asked. What was the owner's solution?

4.14. Blind separation. There are over 500 coins on a table top. You know that exactly 247 coins show heads, and the rest show tails. How can you separate the coins into two groups, each with the same number of heads, *with your eyes closed?*

4.15. ◇ **Joining links**. Given five chains with three links in each chain, what is the least expensive way to create a single chain with fifteen links? Assume that cutting and welding a link costs $3.

4.16. Mending a bracelet. A broken bracelet consists of four pairs of links. It costs $10 to cut and weld a link. What is the least expensive way to create a closed bracelet with eight links?

4.17. ◇ **Space travel**. A new space station is located five days from Earth, with four intermediate fuel stops, one at the end of each day's journey. A space shuttle can carry three days' worth of fuel. There is unlimited fuel on Earth. When the shuttle begins its journey to the space station, all fuel stops are empty. What is the minimum number of days required for the shuttle to ferry fuel to the fuel stops and reach the space station? Describe the daily itinerary of the shuttle.

4.18. Winning strategy. Two players alternately place identical poker chips on a large flat plate. Chips may not overlap. The last player to place a chip entirely on the plate loses. What is a winning strategy for the player who moves first?

4.19. ◇ **Another Lewis Carroll basket problem**. Referring to Example 4.2 for the problem description, assume that in addition to the queen (who weighs 78 kilograms), her daughter (who weighs 42 kilograms), and her son (who weighs 36 kilograms), a 24-kilogram pig, an 18-kilogram dog, and a 12-kilogram cat must also escape. The weight difference between the two baskets cannot exceed 6 kilograms. There must always be a human at each end to help the animals in and out of the baskets. Describe how the queen, her children, and the three animals can escape to the ground in the fewest number of steps. [10]

4.2 Hints and Answers

4.1. (i) HINT: The man can take the wolf, goat, or cabbage on the return trip.
ANSWER: Seven crossings are needed.

(ii) HINT: The brides must be the boat drivers.
ANSWER: Nine crossings are needed.

4.2. (i) HINT: Look for an algebraic solution as in Example 4.3.
ANSWER: One possibility is that the 4-gallon jug is emptied three times and the 7-gallon jug is filled twice.

(ii) HINT: Look for an algebraic solution as in Example 4.3.
ANSWER: One possibility is that the 7-gallon jug is filled four times and the 13-gallon jug is emptied twice.

(iii) HINT: Look for an algebraic solution as in Example 4.3.

ANSWER: One possibility is that the 7-gallon jug is filled five times and the 11-gallon jug is emptied three times.

(iv) HINT: The amount of wine in the three jugs must always be eight quarts.
ANSWER: Seven transfers are needed.

(v) HINT: Use a start-up period before the 9-minute interval begins.

(vi) HINT: Put one hourglass in a rest position on its side.

4.3. (i) HINT: How do you remove half of the water in a cylinder?
ANSWER: Tilt the cylinder until the water line bisects the rectangular cross section of the cylinder.

(ii) HINT: Use the result of part (i).
ANSWER: Six transfers are needed.

4.4. (i) HINT: Start with three coins on each pan.

(ii) HINT: Start with four coins on each pan.

(iii) HINT: Divide the original 80 coins roughly into thirds.

(iv) HINT: What is the worst possible case for the third weighing?

(v) HINT: Start with one coin on each pan for the first weighing. Then add a coin to each pan for the second weighing. Not all logical possibilities can occur.

(vi) HINT: Start with three coins on each pan. Use coins that are known to be good in the second and/or third weighings.

(vii) HINT: Start with four coins in each pan.

(viii) HINT: Weigh selected pairs of coins against each other.

(ix) HINT: If you choose one coin as a test coin and weigh the other coins against it, you need at most 23 weighings. Can you do better using a pair of test coins? You do not need to identify the steel and brass coins; you need only to count them.
ANSWER: At most 13 weighings are needed.

(x) HINT: Start with two coins on each pan. Be patient and organize your work!

4.5. HINT: Weigh three different pairs of coins.

4.6. (i) HINT: Draw a right triangle on a flat surface whose hypotenuse has the same length as the diagonal of the brick.

(ii) HINT: Stack the three bricks in an appropriate way.

4.7. HINT: Use all of your senses.

4.8. HINT: In any five-minute interval, eight blacksmiths can put on eight shoes. Eight horses can be shoed in 20 minutes by 8 blacksmiths. But it takes another 20 minutes for two blacksmiths to shoe the other two horses. Can it be done more quickly?
ANSWER: Yes, the job can be done in 25 minutes.

4.9. HINT: Move the three corner coins.

4.10. HINT: What is the fewest number of ounces that can be distinguished by the scale?

4.11. HINT: Both pairs of gloves can be worn at the same time.

4.12. HINT: Clearly 20 minutes suffice. Can it be done in less?
ANSWER: Yes, it can be done in 15 minutes.

4.13. HINT: Design a box that will hold the pole and comply with the requirements.

4.14. HINT: Read the goal of the problem carefully. Remember that you cannot distinguish heads from tails. How many coins should you turn over?

4.15. HINT: Use a single cut link to join two chains.
ANSWER: The cost is $9.

4.16. HINT: Use a single cut link to join two chains.
ANSWER: The cost is $30.

4.17. HINT: How much fuel must be at each fuel stop when the shuttle begins its final trip to the space station?
ANSWER: The trip can be done in fifteen days.

4.18. HINT: There is only one distinctive place for the first person to place a chip.

4.19. HINT: Start as in Example 4.2, and note that the penultimate step must have the rock on the ground and the son in the tower.
ANSWER: It can be done in 12 steps.

Chapter 5

Parts of the Whole

> *To describe mathematics as only a method of inquiry is to describe da Vinci's "Last Supper" as an organization of paint on canvas. Mathematics is also a creative endeavor.*
>
> — Morris Kline

In this chapter we focus on problems that involve proportions, fractions, and percentages. While these ideas may seem rather elementary in their own right, the chapter will prove that they involve plenty of subtleties.

5.1 Basic Uses of Percentages

Percentages are used for three different purposes: to express fractions, to describe change, and to make comparisons. The first use is quite familiar. For example, if 41.6% of the 625 people at a conference are women, we conclude that $0.416 \times 625 = 260$ of the people at the conference are women. Not quite so obvious is the *inverse* question: If the 5250 men in a town comprise 42% of the total population, the total population is $\frac{5250}{0.42} = 12,500$ (which can be checked by verifying that $0.42 \times 12,500 = 5250$).

To use percentages to describe change, we generally identify a previous value and a current value of a particular quantity. We then express the difference in the two values as a fraction of the *previous value*. For example, the world population was 2.6 billion in 1950 and 6.0 billion in 2000. The percentage change over this 50-year period was

$$\frac{6.0 - 2.6}{2.6} = 1.3 = 130\%.$$

We can describe the change by saying that

- world population increased by 130% (meaning the *change* in population was 130% of the 1950 population), or

- world population increased by a factor of 230% (meaning that the 2000 population is 2.3 times the 1950 population).

Using percentages for comparisons is fraught with at least verbal subtleties. It is essential to identify a *reference quantity*, the quantity to which the comparison is made. For example, in 2000, the per capita income was \$40,640 in Connecticut (highest in the country) and \$20,993 in Mississippi (lowest in the country). To compare the Connecticut income to the Mississippi income, we can compute the percent difference using Mississippi as the reference quantity:

$$\frac{\$40,640 - \$20,993}{\$20,993} = 0.936 = 93.6\%.$$

The correct statement that follows is that the per capita income in Connecticut was 93.6% more than the per capita income in Mississippi, where the phrase *more than* identifies Mississippi as the reference quantity. Equivalently, the per capita income in Connecticut was 1.936 times the per capita income in Mississippi.

We can also compare the per capita income in Mississippi to the per capita income in Connecticut, which means that Connecticut is the reference quantity. Now the relative difference is

$$\frac{\$20,993 - \$40,640}{\$40,640} = -0.483 = -48.3\%.$$

The correct statement is that the per capita income in Mississippi was 48.3% less than the per capita income in Connecticut, where the phrase *less than* identifies Connecticut as the reference quantity. Equivalently, the per capita income in Mississippi was $1 - 0.483 = 0.517$ times the per capita income in Connecticut. Notice that different reference quantities give very different numerical results.

Mathematically, percentages for change and percentages for comparisons are identical. We have distinguished the cases because the corresponding language differs.

In general, if A is $p\%$ more than B, then A is $(100 + p)\%$ of B. If A is $p\%$ less than B, then A is $(100 - p)\%$ of B. If A is $p\%$ more than B, then B is never $p\%$ less than A.

Example 5.1 College tuition. A recent news article states that "the University of North Carolina at Chapel Hill approved a 21% increase for in-state tuition to \$2814." What is the current in-state tuition?

Solution: The key observation is that the new tuition is 21% more than the current tuition; equivalently, the new tuition is 1.21 times the current tuition. Therefore, the current tuition is $\frac{\$2814}{1.21} = \$2326$. It is important to check that a 21% increase from \$2326 is indeed \$2814.

Caution: It is a common mistake to solve this and similar problems by claiming that the current tuition is 21% *less than* the new tuition. This approach gives a current tuition of $0.79 \times \$2814 = \$2223$, which, when increased by 21%, does not equal \$2814. □

Example 5.2 Common errors. Critique the following statements:

(i) If you receive a 10% pay cut, followed a few months later by a 10% pay raise, your salary is unchanged.

(ii) Three successive 5% tax increases results in a total tax increase of 15%.

(iii) Thirty percent of the townspeople are Republican, and 30% of the townspeople are men; therefore $30\% \times 30\% = 9\%$ of the townspeople are Republican men.

(iv) Thirty percent of the townspeople are Republican, and 30% of the Republicans are men; therefore $30\% \times 30\% = 9\%$ of the townspeople are Republican men.

(v) A basketball player hits 85% of her free throws in the first half of a game and 95% of her free throws in the second half of a game. It follows that her free throw percentage for the entire game is 90%.

Solution:

(i) The statement is false. Suppose that your monthly salary is $1000. A 10% pay cut reduces your salary to $900 per month. A subsequent 10% pay raise increases your salary to $990 per month. Your new salary is actually $1.10 \times 0.9 = 0.99$ times your previous salary.

(ii) The statement is false. After three years, taxes will be $1.05^3 = 1.1576$ times current taxes, which amounts to nearly a 16% increase. This problem exhibits the phenomenon of *compounding*.

(iii) The statement is false in general, because it doesn't give the relationship between the set of men and the set of Republicans. The situation in which none of the men is a Republican is consistent with the given facts.

(iv) This statement is true, as it does give the relationship between the set of men and the set of Republicans.

(v) The statement is false unless she had the same number of free throws in both halves. *In general, you cannot average percentages.* ▢

Percentage change problems can be subtle when several changing variables are combined. Here is a modest example [26]. Suppose that the population of a country increased by 1% last year and that the average income per person increased by 2% last year. What was the percentage change in the total national income?

To investigate this question, let last year's population be p and last year's average income be a. Then the total national income was $I = pa$. During the year, all three variables changed by absolute amounts Δp, Δa, and ΔI, where we are given the percentage (or relative changes) $\Delta p/p = 0.01$ and $\Delta a/a = 0.02$. We are asked to find the percentage change $\Delta I/I$.

After the changes occur, we have

$$(I + \Delta I) = (p + \Delta p)(a + \Delta a) = pa + p\Delta a + a\Delta p + \Delta a\Delta p.$$

After combining terms, using $I = ap$, and dividing through by I, we find that

$$\frac{\Delta I}{I} = \frac{\Delta p}{p} + \frac{\Delta a}{a} + \frac{\Delta a\Delta p}{ap}.$$

In many cases, the changes are small; then products of the changes are extremely small and can be neglected. For example, in the present case, $\Delta a\Delta p/(ap) = 0.01 \cdot 0.02 = 0.0002$, which is much smaller than either $\Delta p/p$ or $\Delta a/a$. Thus to a very good approximation, we have

$$\frac{\Delta I}{I} = \frac{\Delta p}{p} + \frac{\Delta a}{a}.$$

Thus the percentage change in the national income is close to $1\% + 2\% = 3\%$.

The relative or percentage change in a quantity z is often denoted $\hat{z}$ and called a **hat derivative**. We have just shown that, for small changes, if $z = xy$, then $\hat{z} = \hat{x} + \hat{y}$. Using the same sort of argument, several other percentage change results can be proved. Here is a short list of useful results, where k is a constant.

The similarities between hat derivatives and ordinary derivatives are not accidental. You should verify that hat derivatives correspond to logarithmic derivatives; that is, $\hat{z} = (d/dx)\log z$. These four rules follow immediately from this fact.

- If $z = kx$, then $\hat{z} = \hat{x}$.
- If $z = xy$, then $\hat{z} = \hat{x} + \hat{y}$.
- If $z = x^k$, then $\hat{z} = k\hat{x}$.
- If $z = x/y$, then $\hat{z} = \hat{x} - \hat{y}$.

Example 5.3 Wages and earnings. Comparing this year to last year, Jack's (after-tax) earnings increased by 9%, and his total hours worked increased by 5%. However, he also increased the fraction of his wages that he put into savings by 3%. By what percentage did Jack's average *hourly* take-home wage change?

Solution: Clearly, we have several variables that are related and changing simultaneously. The first step is to produce the relationship between the variables. Let E be Jack's total annual earnings, h be the number of hours he worked, s be the fraction of his wages that he takes home (that he does not put into savings), and let w be his average hourly wage after savings. Then we have

$$w = \frac{Es}{h} \quad \text{dollars per hour.}$$

We are given that $\hat{E} = 9\%$, $\hat{s} = -3\%$, and $\hat{h} = 5\%$. (Notice that $\hat{s}$ is negative because the amount he saved increased by 3%, so the amount he took home decreased by 3%.) Using the rules given above for the percentage change in products and quotients, we have

$$\hat{w} = \hat{E} + \hat{s} - \hat{h} = 9\% - 3\% - 5\% = 1\%.$$

To a good approximation (because the relative changes are small), Jack's average hourly take-home wage increased by 1%.   □

5.2 Assorted Problems

In this section we proceed by example and solve several problems whose only common feature is the use of proportions and percentages.

Example 5.4 Simpson's paradox. Consider Table 5.1, which gives eighth-grade mathematics test scores in Nebraska and New Jersey. Could these scores be accurate? Explain. [National Assessment of Educational Progress, 1992; from *Chance*, Spring 1999]

Simpson's paradox is named after the Englishman Edward Simpson who described it in 1951. The idea was first described around 1900 by George Yule, a Scottish statistician.

Table 5.1. *Simpson's paradox with test scores.*

	White	Nonwhite	Overall average
Nebraska	281	250	277
New Jersey	283	252	272

Solution: The paradox is that both white and nonwhite students had higher scores in New Jersey than in Nebraska, and yet, the overall average scores were higher in Nebraska than in New Jersey. The only possible explanation is that the two states had very different proportions of white and nonwhite students. Let p be the fraction of white students in Nebraska, and let q be the fraction of white students in New Jersey. The overall averages in the table imply that

$$281p + 250(1 - p) = 277,$$
$$283q + (1 - q)252 = 272.$$

Solving these two equations gives $p = 0.87$ and $q = 0.65$. We see that 87% of the students in Nebraska were white, and they scored only slightly lower than the 65% of white students in New Jersey. At the same time, 35% of the students in New Jersey were nonwhite, and they scored only slightly higher than the 13% of nonwhite students in Nebraska.   □

Example 5.5 Polygraphs. Suppose that a polygraph (lie detector test) is given to 1000 employees, of whom 10 are known to use drugs on the job. When asked, all employees deny using drugs on the job. Assume that the polygraph (like most) is 90% accurate. How many employees are falsely accused of drug use based on the results of the polygraph?

Solution: Table 5.2 shows the number of employees in each of four possible categories. Notice that of the 10 users, 90%, or 9 employees, are correctly identified as users (true positives), and 10%, or 1 employee, is wrongly identified as a nonuser (false negative). Of the 990 nonusers, 90%, or 891 employees, are correctly identified as nonusers (true negatives), and 10%, or 99 employees, are wrongly identified as users (false positives). Thus, of the $99 + 9 = 108$ employees identified as users, 99 employees are wrongly identified as users. If the company fires all employees identified as users, then $\frac{99}{108} = 92\%$ are wrongly fired. This surprising result occurs, despite the high accuracy of the test, because of the small percentage of users.   □

Table 5.2. *Data for polygraph test.*

	Test correct	Test wrong	Totals
Actual user	9	1	10
Actual nonuser	891	99	990
Totals	900	100	1000

Example 5.6 Percentages as probabilities. Table 5.3 shows the results of mammograms for 10,000 women with breast tumors. The base incidence rate for malignant tumors is 0.01, which means that of 10,000 women with breast tumors, 100 have a malignant tumor. The accuracy rate for the mammogram procedure is 85%.

Table 5.3. *Data for 10,000 mammograms.*

	Tumor malignant	Tumor benign	Totals
Positive mammogram	85	1485	1570
Negative mammogram	15	8415	8430
Totals	100	9900	10,000

(i) Check that the numbers in the table are consistent with the base incidence rate and the accuracy of the test.

(ii) What is the probability that a women with a malignant tumor has a positive mammogram?

(iii) What is the probability that a women with a positive mammogram has a malignant tumor?

(iv) Interpret these results.

Solution:

(i) Of the 10,000 women in the sample, 100, or 1%, have a malignant tumor, which is consistent with the base incidence rate. Of the 100 women with malignant tumors, 85%, or 85 women, are correctly diagnosed (true positives), and 15%, or 15 women, are wrongly diagnosed (false negatives). Of the 9900 women with benign tumors, 85%, or 8415 women, have negative tests (true negatives), and 15%, or 1485 women, have positive tests (false positives).

(ii) Of the 100 women with a malignant tumor, 85 have a positive mammogram. Thus, the relative frequency of positive mammograms among women with a malignant tumor is 0.85. Because a relative frequency can be interpreted as a probability, the probability that a women with a malignant tumor has a positive mammogram is 0.85, which is the accuracy of the test.

(iii) Of the 1570 women with a positive mammogram, 85 have a malignant tumor. Thus, the relative frequency of malignant tumors among women with a positive mammogram is $\frac{85}{1570} = 0.054$. Thus, the probability that a women with a positive mammogram has a malignant tumor is 0.054.

(iv) It is easy to confuse the probabilities in parts (ii) and (iii). But the difference is crucial—and not just for mammograms, but for many disease tests. If a woman is known only to have a breast tumor, then the best that can be said is that there is a 0.01 probability that it is malignant (the base incidence rate). If she subsequently has a positive mammogram, then the probability of a malignant tumor, given that the mammogram is positive, is 0.054, as calculated in part (iii). We see that the probability increases only slightly with a positive mammogram, whereas, if the probability of part (ii) is mistakenly used, the outlook is far more pessimistic.   □

5.3 Exercises

5.1. ◇ **Income taxes**. Suppose your net (after-tax) income for the year is $34,500 and that you pay 27% of your income in taxes. What is your gross (before-tax) income?

5.2. Restaurant tips. Anne took three friends out to dinner and paid a total bill of $54.30, which included 6% in taxes and a 15% tip (on the cost of the meals). What was the cost of the meals (before tax and tip)?

5.3. ◇ **Sales tax**. Jack paid a total of $275.78, which included a 7.6% sales tax, for a new CD player that was marked down by 25% from its original price. What was the original retail price of the CD player?

5.4. Smoking rates. An article in the April 20, 1997, *New York Times* states that "the smoking rate among tenth graders jumped 45% to 18.3%." What was the smoking rate among tenth graders before the jump?

5.5. ◇ **Fractional pricing.** Perfume sells for either $30 for 1/60 ounce or $60 for 1/30 ounce. What should the price be for 1/45 ounce?

5.6. Pay raises and cuts. Abe and Zack work together at a slingshot factory for the same wages. One day their boss gives Abe a 10% pay raise and Zack a 10% pay cut. A week later, the boss gives Abe a 10% pay cut and Zack a 10% pay raise. Who earns more now?

5.7. ◇ **Grades.** The final grade in your economics course is determined according to the following weights: assignments 55%, project 15%, midterm 15%, final exam 15%. Going into the final exam you have 185 out of 200 points on the assignments, 45 out of 50 points on the project, and 81 out of 100 points on the midterm. What score (in percentage terms) must you earn on the final exam to have a 90% average for the course?

5.8. Either and/or both. Each child at a school plays either soccer or tennis or both. One-seventh of the soccer players also play tennis. One-ninth of the tennis players also play soccer. Find the fraction of students who play soccer only, tennis only, and both soccer and tennis.

5.9. ◇ **Discount coupons.** Suppose you have three coupons for a local ice cream store. One is for a 25% discount, one is for a 35% discount, and one is for a 40% discount. You decide to take your friends out for ice cream, expecting to get a discount of 25% + 35% + 40% = 100% (no cost!). Although the ice cream store manager is willing to honor each coupon independently, he thinks your discount should be less. Who is correct, and what is the actual discount (as a percentage)?

5.10. Fighting inflation. Complete the following sentences:

(i) Suppose that prices increase between 2004 and 2005 by 4% due to inflation. A computer monitor costs $500 in 2005. Its price in 2004 dollars is _____.

(ii) Assuming a p% annual inflation rate, the purchasing power of my money today is a fraction _____ of what it was last year.

(iii) Assuming a p% annual inflation rate, the purchasing power of my money today has decreased by a fraction _____ in one year.

5.11. ◇ **Batting averages.** At the end of the season, two sluggers, Sammy and Barry, have the first-half, second-half, and overall batting averages shown in the table below. (Batting average is the ratio of number of hits to number of at-bats.) Notice that Sammy has the higher average in the first half and in the second half, and yet he has a lower

overall average. Explain how the figures can be correct by finding the percentage of at-bats each player had in the first and second halves of the season.

	First half	Second half	Overall average
Sammy	0.333	0.500	0.386
Barry	0.200	0.467	0.400

5.12. Water in potatoes. One hundred pounds of potatoes, which are 99% water by weight, are allowed to dry out until they are 98% water by weight. How much do the potatoes now weigh?

5.13. ◇ **Price reduction.** On July 31, the price of a pound of chocolates was $2.56. The next day (August 1), the price was increased, and as a result, sales for the month of August (number of pounds of chocolate sold) were down 23.2% from the previous month, and the total income from sales was down 12.4% for the month. What was the price of chocolates in August? [40]

5.14. Russian elections. According to the *New York Times* (July 5, 1996), Boris Yeltsin won the Russian presidential run-off election by 13 percentage points and had 33% more votes than Gennady Zyuganov. What percentage of the votes did the two candidates receive? Is this result consistent with the fact that the election was a run-off? Explain.

5.15. ◇ **How many voted?** Among those voting in the election, the proportion of men to women was 17 : 15. Had 90 fewer men and 80 fewer women voted, the proportion would have been 8 : 7. How many people voted? [40]

5.16. Labor force statistics. In 2000, there were 65.1 million women employed in the U.S. civilian labor force; that number represents 46.0% of the entire civilian labor force and 60.2% of the employable female population. Assuming that the number of employable women equals the number of employable men, how many men and women were employed and unemployed in 2000?

5.17. ◇ **Disease test.** A disease test that is 80% accurate for both true positives and true negatives is given to 4000 people. The incidence rate of the disease is 1.5%.

(i) Find the number of true positives, false positives, true negatives, and false negatives produced by the test.

(ii) What is the probability that a person with the disease has a positive test? (Equivalently, what fraction of the people with the disease has a positive test?)

(iii) What is the probability that a person with a positive test has the disease?

(iv) Which of these two probabilities should be quoted to a patient who has just tested positive for the disease? Explain.

5.18. Hiring policies [From "Ask Marilyn" in *Parade Magazine*, April 28, 1996]. A company opened a new factory and needed to fill 455 jobs. For the white-collar positions, 200 men and 200 women applied. Twenty percent of the women who applied were hired, while 15% of the men who applied were hired. For the blue-collar positions, 400 men and 100 women applied. Eighty-five percent of the women who applied were hired, while 75% of the men who applied were hired.

(i) Based on the white-collar and blue-collar positions separately, does there appear to be a preference for hiring men or women?

(ii) Based on the overall hiring for all positions, does there appear to be a preference for hiring men or women?

(iii) Give an explanation for the apparent paradox.

5.19. ◇ Staff demographics. On the staff of a company, there are 15 Americans and 24 women. Eight of the employees are neither American nor Swedish, and half of these are women.

(i) Are there more American male employees than Swedish women employees?

(ii) If at least one American man works for the company, could there be fewer than 30 employees? [40]

5.20. Mixtures. When one ounce of water is added to a mixture of acid and water, the new mixture is 20% acid. When one ounce of acid is added to that new mixture, the result is $33\frac{1}{3}$% acid. What is the percentage of acid in the original mixture? [24]

5.21. ◇ More mixtures. Two identical jars are filled with alcohol solutions. The ratio of alcohol to water by volume in one jar is $p : 1$ and in the other is $q : 1$. If the contents of the two jars are combined in a larger jar, what is the alcohol-to-water ratio by volume of the new mixture, in terms of p and q? [24]

5.22. Wages and inflation. This year Flo earned 12% more due to a raise and worked 8% more hours than last year. The inflation rate between last year and this year was 5%.

(i) By what percentage did Flo's real (inflation-adjusted) earnings change from last year to this year?

(ii) By what percentage did Flo's real (inflation-adjusted) average hourly wage change from last year to this year? [26]

5.23. ◇ Student fees. A university increased the fees charged to undergraduates by 7% over last year. At the same time, the number of undergraduates increased by 1%, and the fraction of fees paid personally (out-of-pocket) by undergraduates decreased from 60% to 57% (other sources such as loans and scholarships covered the remainder). By what percentage did the total fees paid personally by undergraduates change? [26]

5.24. Chelsea pensioners (an old classic). In the Chelsea home for pensioners, 70% of the pensioners have lost an eye, 75% have lost an ear, 80% have lost an arm, and 85% have lost a leg. Find the minimum percentage and maximum percentage of pensioners who have lost an eye, an ear, an arm, *and* a leg.

5.25. ◇ Demographics. Of the 100 people attending the news conference, 75 were women, 65 were from Alaska, 90 arrived by bus, and 45 wore gloves. Find the minimum number and maximum number of Alaskan women wearing gloves and arriving by bus at the conference.

5.26. Sale price A TV was marked down by the same percentage in each of three years. After two years, the price of the TV was 51% off the original price. After three years, was the price more than two-thirds off the original price? Explain. [40]

5.27. ◇ Drug trials [From "Ask Marilyn" in *Parade Magazine*, April 28, 1996]. A pharmaceutical company runs two trials of two treatments for a disease. In the first trial, Treatment A cures 20% of the cases, and Treatment B cures 15% of the cases. In the second trial, Treatment A cures 85% of the cases, and Treatment B cures 75% of the cases. And yet, taken over both trials, Treatment A has a 41.7% cure rate, and Treatment B has a 55.0% cure rate. Determine (i) the relationship between the number of patients receiving Treatment A in the two trials and (ii) the relationship between the number of people receiving Treatment B in the two trials that would explain these figures.

5.28. Minimum employees. The percentage of women in an office is (strictly) between 60% and 65%. What is the fewest number of women in the office? [40]

5.29. ◇ Marriage fractions. Three-tenths of all men in a town are married, and two-fifths of all women in the town are married. Assume that all spouses live in the same town and have only one spouse. What fraction of the townspeople are single men? [39]

5.30. More marriage fractions. In a town with 50,000 people, 42% of the men and 28% of the women are married. Assuming no polygamy and assuming that all spouses live in the same town, how many men are in the town? [14]

5.31. ◇ Batting averages. A baseball player currently has an average of 0.350. At the next at-bat, does his average decrease if he fails to get a hit more than it increases if he gets a hit? Does the answer depend on the current batting average? Does the answer depend on the number of at-bats?

Explain. (Batting average is the number of hits divided by the number of at-bats.)

5.32. More batting averages. After getting a base hit, a baseball player noticed that his batting average went up exactly ten points. If this was not his first hit, how many hits does he now have? (Batting average is defined as number of hits divided by the number of at-bats; ten points correspond to $\frac{10}{1000} = 0.01$.) [19]

5.4 Hints and Answers

5.1. HINT: After-tax income is 27% less than before-tax income.
ANSWER: Before-tax income is $47,260.

5.2. HINT: The total bill is 21% more than the cost of the meals.
ANSWER: The cost of the meals is $44.88.

5.3. HINT: First the item is marked down, then it is purchased with sales tax added.
ANSWER: The original retail price was $341.73.

5.4. HINT: Percentages are used in two different ways.
ANSWER: The smoking rate before the increase was 12.6%.

5.5. HINT: Find the price per ounce for 1/60 and 1/30 ounce.
ANSWER: The price for 1/45 ounce should be $40.

5.6. HINT: Calculate their new wages.
ANSWER: Abe's and Zack's new salaries are the same.

5.7. HINT: Write out the expression for the final weighted average, assuming that you know the score on the final exam.
ANSWER: You must have a score of 89.83%.

5.8. HINT: Express the number of children in each category in terms of the number of children who play both sports.
ANSWER: The fractions of children who play soccer only, tennis only, and both sports are $\frac{7}{17}$, $\frac{9}{17}$, and $\frac{1}{17}$, respectively.

5.9. HINT: Apply the discounts successively.
ANSWER: The total discount is 71%.

5.10. HINT: Find the amount that the value of the dollar decreases if prices increase by $p\%$.
ANSWER: (i) The price in 2004 dollars is $480.77. (ii) The purchasing power of a dollar is $\frac{1}{1+p}$ of what it was last year. (iii) The purchasing power of a dollar is $\frac{p}{1+p}$ of what it was last year.

5.11. HINT: Write the overall averages as a weighted average of the first-half and second-half averages.
ANSWER: Sammy had 0.683 of his at-bats in the first half. Barry had 0.251 of his at-bats in the first half.

5.12. HINT: How much "solid potato" is in the batch?
ANSWER: The potatoes weigh 50 pounds.

5.13. HINT: Write August sales as a percentage of July sales and write August income as a percentage of July income.
ANSWER: The August price was $2.92 per pound.

5.14. HINT: Compare the votes received in terms of percentages and percentage points.
ANSWER: Yeltsin received 52.4% of the vote, and Zyuganov received 39.4% of the vote. If it were a run-off election between two candidates, the votes should total 100%.

5.15. HINT: Let the number of men and women be M and W, respectively, and set up the indicated ratios.
ANSWER: There were 170 voting men and 150 voting women.

5.16. HINT: There are four categories within the population of employable Americans: employed/unemployed women/men.
ANSWER: There were 65.1 million employed women, 76.5 million employed men, 43.0 million unemployed women, and 31.8 million unemployed men.

5.17. HINT: First use the incidence rate to find the total number of people with and without the disease. Probability can be interpreted as the fraction of people in the appropriate category.

(i) ANSWER:

	Disease	No disease	Totals
Test positive	48	788	836
Test negative	12	3152	3164
Totals	60	3940	4000

(ii) ANSWER: The probability that a person with the disease tests positive is 0.8.

(iii) ANSWER: The probability that a person who tests positive has the disease is 0.057.

(iv) ANSWER: A patient who has tested positive should be told the probability that, having tested positive, s/he has the disease (which is 0.057).

5.18. HINT: The hiring rates are given for the white- and blue-collar positions separately. Compute the hiring rates for the categories combined.

(i) ANSWER: There appears to be a preference for hiring women in each category separately.

(ii) ANSWER: Combining the categories, the hiring rate for men is 55.0%, while the hiring rate for women is 41.7%.

(iii) ANSWER: While the hiring rate for women is greater in both categories individually, the pool of men for blue-collar jobs is four times larger than the pool of women. As a result, almost three times as many men as women were hired into blue-collar jobs. This difference is sufficiently large that the overall hiring rate for men is greater than the overall hiring rate for women.

5.19. HINT: A table helps organize the information.
ANSWER: (i) There are five more Swedish women than American men. (ii) With at least one American man, there could be as few as 29 employees.

5.20. HINT: Let a and w be the number of ounces of acid and water, respectively, in the original solution.
ANSWER: The original solution has 25% acid.

5.21. HINT: Calculate the amount of alcohol and water in each jar, assuming that the volumes of solution in each jar are equal.
ANSWER: The ratio of alcohol to water by volume is $\frac{p+q+2pq}{p+q+2}$.

5.22. HINT: Refer to Exercise 5.10 for the effects of inflation. You can use hat derivatives, as in Example 4.3, or calculate the result exactly.
ANSWER: (i) Flo's inflation-adjusted earnings increased by 6.7% (or 7% using hat derivatives). (ii) Flo's hourly wage decreased by 0.99% (or 1% using hat derivatives).

5.23. HINT: Be sure to compute the percent change in the fraction of fees paid personally.
ANSWER: The fees collected directly from students increased by 2.6% (or 3% using hat derivatives).

5.24. HINT: A picture might help determine the minimum and maximum overlap among the injuries.

ANSWER: A maximum of 70% of the pensioners have all four injuries. A minimum of 10% of the pensioners have all four injuries.

5.25. HINT: A picture might help determine the minimum and maximum overlap between the properties.
ANSWER: A maximum of 45 people fit the description. A minimum of zero people fit the description.

5.26. HINT: Watch the wording. The percentage *off* the original price is different than the percentage *of* the original price.
ANSWER: After three years, the price is less than two-thirds off the original price.

5.27. HINT: Let a_1 and a_2 be the number of patients receiving Treatment A in the first and second trials, respectively.
ANSWER: The given figures result if twice as many people receive Treatment A in trial 1 as in trial 2, and if twice as many people receive Treatment B in trial 2 as in trial 1.

5.28. HINT: Look carefully for integer values for the number of women and the number of employees that satisfy the given conditions.
ANSWER: With five women and eight employees, 62.5% of the employees are women.

5.29. HINT: The number of married men equals the number of married women!
ANSWER: Two-fifths of the townspeople are unmarried men.

5.30. HINT: The number of married men equals the number of married women!
ANSWER: There are 20,000 men in the town.

5.31. HINT: Compute the new batting average with and without a hit at the next at-bat.
ANSWER: The average increases more with a hit than it decreases without a hit. This conclusion depends on the current average, but not on the number of hits.

5.32. HINT: Compute the batting average before and after the hit.
ANSWER: The player now has 19 hits.

Chapter 6

A World of Change

Nothing in the world lasts save eternal change.
—Honorat de Bueil, 1600

All things must change to something new, to something strange.
—Henry Wadsworth Longfellow

The term *rate* has many uses and meanings. Hikers refer to the rate of ascent of a trail (for example, 800 vertical feet per mile). Epidemiologists measure death rates (for example, the death rate for cancer in the United States is about 200 deaths per 100,000 people per year). Demographers estimate population growth rates (for example, world population is increasing at a rate of 1.2% per year). Ecologists measure the rate of consumption of resources (for example, the Amazon rain forest is depleted at a rate of 1000 acres per day). Economists use the consumer price index to chart the annual rate of change of prices. The common feature in most uses of *rate* is that the term describes how one variable changes with respect to another.

Mathematically, the crucial distinction in rate problems is between *constant rate* problems, which can be solved using algebra, and *variable rate* problems, which generally require the use of calculus. We consider constant rate problems in this chapter, and deal with variable rate problems in the next chapter.

Consider a quantity Q that changes at a constant rate R with respect to time. Then R will have units of Q per unit time. For example, if Q is distance, then R is a speed, and R might have units of meters per second. If Q is a volume of water, then R is a flow rate, and it might have units of liters per second. The fundamental rule in working with constant rates is

Amount of Q accumulated = rate × time elapsed   or $Q = Rt$.

In other words, constant rate implies that Q changes linearly in time.

If the rate is a speed, then this rule is the familiar *distance equals speed times time* formula. However if, R is a flow rate, then the rule gives the total amount of material that flows or accumulates in a given time. If R is a population growth rate, then the rule gives the number of people added to the existing population in a given time. This rule is fundamental in the following examples and problems.

Example 6.1 (Not) averaging rates. This example illustrates an important and easily forgotten point. Suppose you ride your bike up a hill at a constant speed of 8 miles per hour, turn around, and descend the same hill at a constant speed of 18 miles per hour. What is your average speed for the round trip?

Solution: The temptation may be to average the ascent speed and the descent speed, and conclude that the average round-trip speed is 13 miles per hour. However, this argument fails to recognize that more time is spent riding uphill than riding downhill. The lesson of this example is that *one must be careful about averaging rates.*

The correct approach is to find the total distance traveled and divide it by the total time elapsed. Suppose the hill is L miles long. Then the total length of the trip is $2L$ miles and the total time elapsed is $L/8$ hours for the ascent plus $L/18$ hours for the descent. Thus, the average speed for the round trip is

$$\overline{v} = \frac{2L}{\frac{L}{8} + \frac{L}{18}} = 11.08 \ \frac{\text{mi}}{\text{hr}}.$$

Because more time is spent at the lower speed, the average round-trip speed is less than the average of the two speeds. Notice that the result is independent of the length of the trip.

In general, if the first half of the trip is done at a speed v_1 and the second half of the trip is done at a speed v_2, the round-trip average speed is

$$\overline{v} = \frac{2}{\frac{1}{v_1} + \frac{1}{v_2}} = \frac{2v_1 v_2}{v_1 + v_2}.$$

(In the special case that $v_1 = v_2 = v$, we have $\overline{v} = v$, as expected.) This expression says that $\overline{v}$ is the **harmonic mean** of v_1 and v_2. In general, we can show quite elegantly that $\overline{v}$ is *always* less than or equal to the **arithmetic mean** of the two speeds; that is,

$$\overline{v} \le v_a = \frac{1}{2}(v_1 + v_2).$$

Recall that $v_g = \sqrt{v_1 v_2}$ is the **geometric mean** of the two speeds and that $v_h = 2/(v_1^{-1} + v_2^{-1})$ is the harmonic mean of the two speeds. We see that

To show that the geometric mean of two positive numbers is less than or equal to their arithmetic mean, start with the inequality $0 \le (\sqrt{a} - \sqrt{b})^2$ and expand the right-hand side. In general, for any n positive numbers, their harmonic mean is less than or equal to their geometric mean, which is less than or equal to their arithmetic mean.

$$\overline{v} = v_h = \frac{2v_1 v_2}{v_1 + v_2} = \frac{v_g^2}{v_a}.$$

Using a familiar result that the geometric mean of two positive numbers never exceeds their arithmetic mean ($v_g \le v_a$), we have

$$\overline{v} = v_h = \frac{v_g^2}{v_a} \le \frac{v_a^2}{v_a} = v_a,$$

with equality only if $v_1 = v_2$. □

Example 6.2 Work rates. Suppose that when Tracy and Sue work together, they can complete a job in 36 minutes. When Tracy works alone, she can complete the same job in 60 minutes. How long does it take Sue to complete the job working alone?

Solution: The assumption in problems of this sort is that when working together, Tracy and Sue work independently and work at the same rate as they would when working alone; in other words, there is no teamwork. The relevant rate in this problem is *jobs per minute*. If we let Tracy's work rate be r_T, then we can write

$$r_T = \frac{1}{60}\frac{\text{job}}{\text{minute}} \quad \text{or} \quad \frac{1}{r_T} = 60\frac{\text{minutes}}{\text{job}}.$$

We see that r_T is a work rate, and its reciprocal, r_T^{-1}, is the time required to complete one job.

The assumption about independent work means that when Tracy and Sue work together, their combined rate, $\frac{1}{36}$ jobs per minute, is the sum of their individual rates. So we have

$$r_T + r_S = \frac{1}{36}\frac{\text{jobs}}{\text{minute}}.$$

Knowing that $r_T = \frac{1}{60}$, we can solve for r_S:

$$r_S = \left(\frac{1}{36} - \frac{1}{60}\right) = \frac{1}{90}\frac{\text{jobs}}{\text{minute}}.$$

This means that Sue needs 90 minutes to do the job working alone.

This sort of problem can be generalized in many ways (see the Exercises section). If n people are working on a job (again, assuming independence of work) and we are given the rates or completion times for n different subgroups of workers (of size 1 to n), then it is possible to find the rates and completion times for all of the workers. ☐

Another concept arising in many rate problems is **relative speed**. The idea arises with boats on streams, planes with tailwinds, or whenever there are two different frames of reference for a moving object. Here is an example.

Example 6.3 Round trip on a river. Sue paddles upstream from the boathouse at a rate of 6 km/hr relative to the current that flows at 2 km/hr. She paddles 3 kilometers, then turns around and paddles back to the boathouse, still at a rate of 6 km/hr relative to the current. How long did she paddle?

Solution: The challenge with this and similar problems is visualizing the motion of the boat—relative to the current and relative to the shore. Sue's rate of $v_c = 6$ km/hr relative to the current is the speed you would measure if you observed Sue while floating downstream with the current. Because she paddles with the same effort going upstream and downstream, her speed relative to the current is the same for the entire trip. However, to find the time required for the upstream and downstream journeys, we need to use her speed relative to the shore, v_s (because the length of the trip is given as distance along the shore). If you were observing from the shore, you would see Sue paddling *against* the current on the upstream trip, so her speed relative to the shore would be

$$v_s = v_c - 2 \text{ km/hr} = 4 \text{ km/hr}.$$

Similarly, if you were observing from the shore, you would see Sue paddling *with* the current on the downstream trip, so her speed relative to the shore would be

$$v_s = v_c + 2 \text{ km/hr} = 8 \text{ km/hr}.$$

The time for the total trip is the sum of the times for the upstream and downstream journeys, or

$$t = \frac{3 \text{ km}}{4 \text{ km/hr}} + \frac{3 \text{ km}}{8 \text{ km/hr}} = \frac{9}{8} \text{ hr.} \qquad \square$$

We close this chapter with an example of an important idea that finds application in many different situations. The basic question is this: Imagine two people running laps on a circular track at different speeds. When and where do they pass each other? While this may sound like a question about track meets, it actually applies to many phenomena that have periodic or cyclic behavior. For example, models of coupled oscillators or biological cycles often use this basic idea. And when the speeds of the runners are variable, the behavior of such models can be truly complex.

Example 6.4 Running laps. Beginning at the same point on a circular track, Polly and Quinn start running in the same direction at the same time. When and where do they pass each other when

(i) Polly runs at 1 lap per minute and Quinn runs at $\frac{3}{4}$ laps per minute, and

(ii) Polly runs a lap in $\frac{3}{2}$ minutes and Quinn runs a lap in 6 minutes?

Solution: It is instructive to formulate a solution in general terms; then we can answer the specific questions. Suppose Polly runs at a rate of p laps per minute, and Quinn runs at a rate of $q < p$ laps per minute. Then Polly's lap count is given by $\theta_p(t) = pt$, and similarly, Quinn's lap count is $\theta_q(t) = qt$, where t is the elapsed time in minutes. Note that when $\theta_p(t)$ or $\theta_q(t)$ has an integer value, it means that Polly or Quinn completes an integer number of laps and passes the starting point.

In problems of this nature, θ_p and θ_q are often called the **phases** of the runners. It is useful to introduce the **phase difference**, which is the separation of the runners, measured in laps; it is given by $\phi(t) = \theta_p(t) - \theta_q(t) = (p-q)t$. Whenever $\phi(t)$ has an integer value, it means that the runners are an integer number of laps apart, which says that one runner passes the other. It is now straightforward to find the times at which one runner passes the other. They are given by $\phi(t) = (p - q)t = k$, where k is a positive integer. The times at which the runners pass each other are

$$t^* = \frac{k}{p - q}, \quad \text{where} \quad k = 1, 2, 3, \ldots .$$

This argument tells us *when* the runners pass each other. Determining *where* they pass each other can be trickier. If $\theta_p(t^*) = m$ and $\theta_q(t^*) = n$, where m and n are integers, then the runners pass each other at the starting point. However, $\theta_p(t^*)$ and $\theta_q(t^*)$ may differ by an integer, but not be integers. In this case, the runners pass each other at a point other than the starting point. Here are three general rules that follow from these observations (see Exercise 6.35):

- If $\frac{p}{p-q}$ is an integer, then the runners will pass each other only at the starting point, every $\frac{p}{p-q}$ laps of the faster runner.

- If $\frac{p}{p-q}$ is rational, but not an integer, then the runners will pass each other at the starting point and at other points along the track.

- If $\frac{p}{q}$ is irrational, the runners will never pass each other at the starting point, but will pass each other at other points along the track.

We can now turn to the questions in the example. In part (i), we have $p = 1$ lap per minute and $q = \frac{3}{4}$ laps per minute. The phases of the runners are $\theta_p(t) = t$ and $\theta_q(t) = \frac{3t}{4}$. The phase difference of the runners is $\phi(t) = (p - q)t = \frac{t}{4}$. We see that the runners pass each other when $\phi(t) = k$, where k is an integer; thus the passing times are $t^* = 4k$. To find the location of the passings, note that $\theta_p(t^*) = 4k$, which is always an integer. As a check, $\theta_q(t^*) = k$, which is always an integer. Thus, the runners pass each other only at the starting point, and do so every fourth lap of Polly.

In part (ii), because Polly runs a lap in $\frac{3}{2}$ minutes, we have $p = \frac{2}{3}$ laps per minute; because Quinn runs a lap in 6 minutes, $q = \frac{1}{6}$ laps per minute. The phases of the runners are $\theta_p(t) = \frac{2t}{3}$ and $\theta_q(t) = \frac{t}{6}$. The phase difference of the runners is $\phi(t) = (p - q)t = \frac{t}{2}$. The runners pass each other when $\phi(t) = k$, where k is an integer; now the passing times are $t^* = 2k$. To find the location of the passings, we need to look at the phase for each runner. For Polly, $\theta_p(t^*) = \frac{4k}{3} = \frac{4}{3}, \frac{8}{3}, 4, \ldots$. For Quinn, $\theta_q(t^*) = \frac{k}{3} = \frac{1}{3}, \frac{2}{3}, 1, \ldots$. Note that the respective lap counts of the runners correspond to the same point on the track. The first passing occurs after Polly runs $1\frac{1}{3}$ laps and Quinn runs $\frac{1}{3}$ laps. The second passing occurs after Polly runs $2\frac{2}{3}$ laps and Quinn runs $\frac{2}{3}$ laps. The third passing occurs at the starting point after Polly runs four laps and Quinn runs one lap. Thus, every third passing occurs at the starting point. □

6.1 Exercises

6.1. ◇ **(Not) averaging rates**. A car went 4 miles up a hill at 10 miles per hour and back down the same hill at 20 miles per hour. What was the average speed for the round trip?

6.2. (Not) averaging rates. A car went 4 miles up a hill at 40 miles per hour and 6 miles down the back of the hill at 70 miles per hour. What was the average speed of the trip?

6.3. ◇ **Tunnel time**. A one-mile-long freight train went through a one-mile-long tunnel at 15 miles per hour. How long did it take the entire train to pass through the tunnel?

6.4. Walking and riding. If a lady walks to work and drives home, the trip takes one and a half hours. When she drives both ways, it takes half an hour. How long would it take her to walk the round trip? [30]

6.5. ◇ **Motorboat in a current**. At full speed, a motorboat can go upstream 10 miles in 15 minutes and downstream 10 miles in 9 minutes. At full speed, how long will it take the boat to go 10 miles with no current?

6.6. Passing time. Two trains are traveling toward each other on parallel tracks. Train A is $\frac{1}{2}$ mile long and travels at 20 miles per hour. Train B is $\frac{1}{3}$ mile long and travels at 30 miles per hour. How long does it take the trains to pass each other completely (from the instant the engines meet to the instant that the cabooses pass each other)?

6.7. Passing trains.

(i) ◇ A train left Denver bound for Omaha at a speed of 80 miles per hour, and at the same time another train left Omaha bound for Denver at a speed of 100 miles per hour. When the trains passed each other, what fraction of its trip had the Denver train completed?

(ii) At midnight a train left Denver bound for Chicago, and another train left Chicago bound for Denver, both traveling at constant speeds on adjacent tracks. The first train took 12 hours to complete the trip, and the second train took 16 hours to complete the trip. At what time did the trains pass each other?

(iii) ◇ A train left Denver bound for Omaha, a distance of 500 miles, at a speed of 80 miles per hour. Two hours later, another train left Omaha bound for Denver at a speed of 100 miles per hour. When the trains passed each other, how far had the Denver train traveled?

(iv) Train A left City A at a constant speed v_A. A distance d away, train B left City B at a constant speed v_B, k hours after train A left. In terms of the variables of the problem, where along the adjacent tracks did the trains pass each other?

(v) ◇ A train left Boston for New York, a distance of 220 miles, at 70 miles per hour. An hour later, a train left New York for Boston at 60 miles per hour. How far apart were the trains one hour before they met?

6.8. Leaky reservoir. Five hundred gallons of water leaked from a reservoir at a constant rate of 8 gallons per day. The leak was partially sealed, and an additional 300 gallons leaked from the reservoir at a constant rate of 5 gallons per day. What was the overall rate of leakage over this time period?

6.9. Work rates.

(i) ◇ Working together (but independently), Arlen, Ben, and Carla can complete a job in one hour. Working alone, Ben and Carla can complete the same job in 2.5 and 3.5 hours, respectively. How long would it take Arlen to complete the same job working alone?

(ii) On average, a hen and a half can lay an egg and a half in a day and a half. On average, how many eggs can nine hens lay in nine days?

(iii) ◇ Twenty people can make 4 hats in 2 hours. How long will it take 15 people to make 30 hats? How many people are needed to make 40 hats in 4 hours? How many hats can be made by 5 people in 12 hours?

(iv) If 5 girls pack 5 boxes in 5 minutes, how many girls are needed to pack 100 boxes in 25 minutes?

(v) ◇ Ann and Betty can do a job in 10 days; Ann and Carol can do the same job in 12 days; Betty and Carol can do the same job in 20 days. How long will it take Carol to do the job alone?

6.10. Machine rates. Working alone, photocopy machine C requires 40 minutes to complete an 800-page job. Working together, machines B and C require 25 minutes for the same 800-page job. With machines A, B, and C working together, the job takes 10 minutes. How long does it take machine A to complete the job, working alone?

6.11. ◇ **Filling a tank.** Pipes A and B can fill a tank in two hours and three hours, respectively. Pipe C can empty the same full tank in five hours. If all pipes are opened at the

same time when the tank is empty, how long will it take to fill the tank?

6.12. Open and shut valves. Each of valves A, B, and C, when open, releases water into a tank at its own constant rate. With all three valves open, the tank fills in 1 hour. With only valves A and C open, it takes 1.5 hours to fill the tank, and with only valves B and C open, it takes 2 hours. How long does it take to fill the tank with valves A and B open? [24]

6.13. ◇ **Filling another tank.** Joe opened two input pipes to a tank, but forgot to close the drain. The tank was half full when he noticed his error and closed the drain. If it takes one input pipe ten hours to fill the tank and the other input pipe eight hours to fill the tank (with the drain closed), and if it takes the drain six hours to empty the tank when it is full (with no input pipes open), how long did it take Joe to fill the tank on this occasion? [17]

6.14. Train schedules. Every hour on the hour a train leaves New York for Philadelphia, and every hour on the half hour a train leaves Philadelphia for New York. If the trip takes exactly two hours in either direction, how many northbound trains will a southbound train meet during the trip (outside of the stations)?

6.15. ◇ **Bike race.** Two cyclists racing on parallel roads maintain constant speeds of 30 mph and 25 mph. The faster cyclist crosses the finish line one hour before the slower cyclist. How long was the race (in miles)?

6.16. Clock hands. What is the first time after 3:00 when the minute hand is as far past the 6 as the hour hand is before the 6? [17]

6.17. ◇ **Burning issue.** Two candles of equal length were lit at the same time. One candle took six hours to burn out, and the other candle took three hours to burn out. After how much time was one candle exactly twice as long as the other candle?

6.18. Dueling candles. Two candles of lengths L and $L+1$ were lit at 6:00 and 4:30, respectively. At 8:30 they had the same length. The longer candle died at 10:30, and the shorter candle died at 10:00. Find L.

6.19. ◇ **Headstart.** Because Boat A travels 1.5 times faster than Boat B, Boat B was given a 1.5-hour headstart in a race. How long did it take Boat A to catch Boat B?

6.20. Raft and canoe. At the same time that Jack began floating downriver in a raft, Joan set out in a canoe half a mile downstream. She paddled downstream then turned around and paddled back upstream, reaching her starting point just as Jack floated by in the raft. Assuming that Joan paddled at a constant speed relative to the current and that her paddling speed in still water is ten times the speed of the current, how far did she paddle? [17]

6.21. ◇ **Up and downwind.** A plane flew into a headwind and made the outbound trip in 84 minutes. It turned around and made the return trip with a tailwind in 9 minutes less than it would have taken with no wind. Assuming that the plane's ground speed and the wind speed are constant, what are the possible times for the return trip? [24]

6.22. Drains and pipes. A large tank is filled at a constant rate by a supply pipe; it is also emptied at a constant rate by ten identical drain pipes. A full tank with Q gallons empties in 2.5 hours when the supply pipe and all ten drain pipes are opened. A full tank empties in 5.5 hours when the supply pipe and six drain pipes are opened. How long does it take a full tank to empty when the supply pipe and three drain pipes are opened? [17]

6.23. ◇ **The escalator.** An escalator with n uniform steps visible at all times descends at constant speed. Two boys, A and Z, walk down the escalator steadily as it moves, A negotiating twice as many escalator steps per minute as Z. A reaches the bottom after taking 27 steps, while Z reaches the bottom after taking 18 steps. Find n. [24]

6.24. A good question. Cars A and B travel the same distance. Car A travels half of the *distance* at u miles per hour and the other half at v miles per hour. Car B travels half of the *time* at u miles per hour and the other half at v miles per hour. Compare the average speeds of the two cars. [24]

6.25. ◇ **An old riddle.** A woman usually takes the 5:30 train home from work, arriving at the station at 6:00, where her husband meets her to drive her home. One day she left work early and took the 5:00 train, arrived at the station at 5:30, and began walking home. Her husband, leaving home at the usual time, met his wife along the way and brought her home ten minutes earlier than usual. How long did the woman walk?

6.26. Ants on a stick. One hundred ants are placed on a stick that is one meter in length. Each ant crawls to either the left or the right with a constant speed of one meter per minute. When two ants meet, they bounce off each other and reverse direction. When an ant reaches an end of the stick, it falls off. What is the maximum amount of time you would need to wait for all of the ants to fall off the stick? [Attributed to Felix Várdy by Francis Su, Harvey Mudd College.]

6.27. ◇ **The bike and the parade.** A cyclist began at the tail of a parade that was four kilometers long and rode in the direction that the parade was moving. By the time the cyclist reached the head of the parade and returned to the tail, the parade had moved six kilometers. Assuming that the cyclist and the parade moved at constant (but different) speeds, how far did the cyclist ride?

6.28. A Lewis Carroll favorite. Some friends left home at 3:00 and walked along a level road, up a hill, back down the same hill, and back home on the same level road without stopping, arriving home at 9:00. Their speed was four miles per hour on the level, three miles per hour on the uphill, and six miles per hour on the downhill. Within half an hour, when did they reach the top of the hill, and what was the total distance they walked? [17]

6.29. ◇ **What's that in the road... ahead?** Bob was traveling 80 km/hr behind a truck traveling 65 km/hr. How far behind the truck was Bob one minute before the crash? [10]

6.30. Rowing in a current. At his usual rate Bernie can row 15 miles downstream in five hours less time than it takes him to row 15 miles upstream. If he doubles his usual rate, his time downstream is only one hour less than the time upstream. What is the rate of the current in miles per hour? [24]

6.31. ◇ **Linked tanks I.** Two tanks are each filled with 90 gallons of water. Tank A has an outflow pipe, which feeds into Tank B, which has an outflow pipe that empties into a drain. The outflow rates of Tanks A and B are 3 gallons per minute and 4 gallons per minute, respectively. If the outflow pipes are opened at the same time, when are both tanks first empty?

6.32. Linked tanks II. Two identical tanks are initially full of water. Tank A has an outflow pipe that feeds into Tank B, which has an outflow pipe that empties into a drain. The outflow rates of Tanks A and B are 3 gallons per minute and 4 gallons per minute, respectively. At a certain time after the pipes are opened, the volume of water in Tank B is twice the volume of water in Tank A. Ten minutes later, the volume of water in Tank B is three times the volume of water in Tank A. At what time is the volume of water in Tank B four times the volume of water in Tank A?

6.33. ◇ **Harvesting.** Alan and Brenda have a vegetable garden and an apple orchard. Working together, they can harvest the garden in 3 hours, whereas Brenda, working alone, needs 12 hours. Together, they can harvest the orchard in 2 hours, whereas Alan, working alone, needs 10 hours. Compare the time needed for the following two strategies: (A) Alan and Brenda work together harvesting the garden and then work together harvesting the orchard, (B) Alan starts working alone in the garden and Brenda starts working alone in the orchard, with the person who finishes first helping the other person. [1]

6.34. Running laps I. Polly and Quinn begin running laps on an oval track in the same direction, starting at the same time at the same point.

(i) If Polly runs a lap in 2 minutes and Quinn runs a lap in 2.5 minutes, when and where will Polly pass Quinn?

(ii) If Polly runs $\frac{5}{3}$ laps per minute and Quinn runs $\frac{2}{3}$ laps per minute, when and where will Polly pass Quinn?

(iii) If Polly runs 5 laps per minute and Quinn runs 4 laps per minute, when and where will Polly pass Quinn if Quinn has a $\frac{1}{3}$-lap headstart?

6.35. ◇ **Running laps II**. Confirm the assertions of Example 6.4: Beginning at the same time and place on a track, and running in the same direction, Polly runs p laps per minute, and Quinn runs $q < p$ laps per minute.

(i) If $\frac{p}{p-q}$ is an integer, then the runners will pass each other only at the starting point every $\frac{p}{p-q}$ laps of Polly.

(ii) If $\frac{p}{p-q}$ is rational, but not an integer, then the runners will pass each other at the starting point and at other points along the track.

(iii) If $\frac{p}{q}$ is irrational, the runners will never pass each other at the starting point, but will pass each other at other points along the track.

6.36. Running opposite laps. Two boys start running in opposite directions from the same point A on a circular track. Their speeds are 9 feet per second and 5 feet per second. If they start at the same time and finish when they first meet again at point A, how many times do they pass each other along the track, excluding the start and finish? [24]

6.37. ◇ **Running alternate laps**. Harry and Skip run laps on a circular track, always starting at the same time and place, and running at constant (but different) speeds. On the first morning, Skip lapped Harry twice and caught him for the third time as Harry was finishing his first lap. On the second day, they ran at the same speeds as the previous day, but in opposite directions. Not counting the start and the finish, how many times did the runners pass each other during Harry's first lap (on the second day)? [40]

6.38. A stagecoach story. A stagecoach left Deadwood for Tombstone at the same time that a stagecoach left Tombstone for Deadwood. They met at a point 24 miles closer to Deadwood than to Tombstone. At this point the drivers switched coaches so they could return to their home towns. The driver from Deadwood completed his trip 9 hours after the switch, and the driver from Tombstone completed her trip 16 hours after the switch. Assuming the *coaches* maintained constant (but different) speeds, what is the distance from Deadwood to Tombstone? [1]

6.39. ◇ **The slow and the swift**. A passenger train travels n times faster than a freight train, and takes n times longer to pass the freight train when the trains are traveling in the same direction as it takes to pass the freight train when they are traveling in opposite directions. What is n? [17]

6.40. Marathon paces. Ann and Bret ran a (26.2-mile) marathon, with Ann running at a constant speed of 8 minutes per mile. Bret ran in fits and starts, but took exactly 8 minutes and 1 second to complete every one-mile interval (that is, every interval from x to $x + 1$ miles, where $0 \leq x \leq 25.2$). Is it possible that Bret finished ahead of Ann? [provided by Stan Wagon, Macalester College]

6.41. ◇ **Feynman's hat**. A man paddled upstream on a river flowing at a rate of 2 km/hr. At 12:00, his hat blew away and floated downstream with the current. Only after paddling 3 km did he realize that he had lost his hat, at which point he turned around and paddled downstream until he overtook his hat. If the man paddled at a constant rate of 6 km/hr relative to the water, when did he overtake his hat? Try a change of perspective.

6.42. Walk/ride strategy. Three men took a trip and used the following scheme. They started at the same time with Paul riding in the car with Bob, while Peter walked. After a while, Bob dropped Paul, who continued on foot, and returned to pick up Peter. Bob and Peter rode in the car until they overtook Paul, at which point they repeated the process. Assume that Peter and Paul walked at 3 miles per hour and that Bob drove the car at 20 miles per hour. What was the average speed of the entire group? [29]

6.43. ◇ **Grazing time**. Gilbert owns a pasture in which the grass grows at a constant rate up to a maximum uniform depth. Gilbert also owns a cow, a horse, and a sheep that all graze in the pasture. Beginning with a pasture full of grass, the grass can feed all three animals for 20 days. It feeds the cow and the horse alone for 25 days, the cow and the sheep alone for $33\frac{1}{3}$ days, and the horse and the sheep alone for 50 days. How long would the grass feed the cow alone? the horse alone? the sheep alone? [1]

6.2   Hints and Answers

6.1. HINT: How long did the car spend on each leg of the trip?
ANSWER: The average speed for the round trip was $\frac{40}{3}$ miles per hour.

6.2. HINT: How long did the car spend on each leg of the trip?
ANSWER: The average speed for the trip was 53.8 miles per hour.

6.3. HINT: How far must the head of the train travel? Draw a picture.

ANSWER: The train takes $\frac{2}{15}$ hours (8 minutes) to pass through the tunnel.

6.4. HINT: Find the walking rate and riding rate.

ANSWER: The woman needs $\frac{5}{2}$ hours to walk both ways.

6.5. HINT: Use the speed of the boat relative to the land on the upstream and downstream trips.

ANSWER: The time required to travel 10 miles with no current is $\frac{45}{4}$ minutes.

6.6. HINT: Consider the problem from the perspective of Train A.

ANSWER: It takes one minute for the trains to pass each other.

6.7. HINT: At the passing point, the times traveled by each train are equal, when adjusted for headstarts.

   (i) ANSWER: The Denver train completed $\frac{4}{9}$ of its trip.

  (ii) ANSWER: Both trains reached the passing point in $\frac{48}{7} \doteq 6.9$ hours.

 (iii) ANSWER: The Denver train traveled 311.1 miles.

 (iv) ANSWER: The trains passed $\frac{dv_A + kv_A v_B}{v_A + v_B}$ miles from Town A.

  (v) ANSWER: The trains were 130 miles apart one hour before they met.

6.8. HINT: Find the times for the 500-gallon and 300-gallon leaks.

ANSWER: The overall leakage rate was 6.53 gallons per day.

6.9. HINT: Find the work rates and add work rates when people work together independently.

   (i) ANSWER: Arlen can do the job in $\frac{35}{11} \doteq 3.2$ hours working alone.

  (ii) ANSWER: The hens can lay 54 eggs.

 (iii) ANSWER: 20 hours, 100 people, 6 hats.

 (iv) ANSWER: Twenty girls are needed.

  (v) ANSWER: It takes Carol 60 days to do the job.

6.10. HINT: Find the work rates of the machines either in jobs per minute or pages per minute.

ANSWER: Machine A needs $\frac{50}{3} \doteq 16.7$ minutes.

6.11. HINT: Find the fill rate of Pipes A and B; find the outflow rate of Pipe C.

ANSWER: The time to fill the tank is $\frac{30}{19} \doteq 1.58$ hours.

6.12. HINT: Find the fill rate of each valve working alone.

ANSWER: The time required for valves A and B to fill the tank is $\frac{6}{5}$ hours.

6.13. HINT: Find the fill rate of Pipes A and B; find the outflow rate of the drain.

ANSWER: The time to fill the tank is 10.8 hours.

6.14. HINT: Suppose the New York train leaves at noon. Find the relevant departure times for the Philadelphia trains.

ANSWER: The New York train will pass four Philadelphia trains.

6.15. HINT: Find the finishing times for both cyclists in terms of the length of the race.

ANSWER: The length of the race is 150 miles.

6.16. HINT: By looking at the clock, the position in question occurs around 3:40.

ANSWER: The position in question occurs at 3:41:30.

6.17. HINT: Find the burning rates of the candles.

ANSWER: After two hours one candle was twice the length of the other candle.

6.18. HINT: Let $t = 0$ correspond to 4:30. Find the length functions for both candles.

ANSWER: The length of the shorter candle is $L = 8$.

6.19. HINT: Equate the travel times from the moment Boat A begins.

ANSWER: It took Boat A three hours to catch Boat B.

6.20. HINT: Use Joan's paddling speed relative to the shore when she paddles upstream and downstream.

ANSWER: Joan traveled 4.95 miles.

6.21. HINT: Show that the plane's speed with no wind is the average of its downwind and upwind speeds.

ANSWER: The return trip took 63 or 12 minutes.

6.22. HINT: Find the net drainage rate with ten and six drain pipes open.

ANSWER: With three drain pipes open, the tank will drain in 55 hours.

6.23. HINT: The number of steps showing, n, is the number of steps taken by A plus the number of escalator steps that have appeared during the descent, with a similar statement for Z.

ANSWER: There are 54 steps showing.

6.24. HINT: Find the distance traveled by each car during the first and second halves of its trip.

ANSWER: The average speed of Car B is never greater than the average speed of Car A.

6.25. HINT: Algebra is not needed. The husband drives for the same amount of time in both directions.

ANSWER: The wife walked for 25 minutes.

6.26. HINT: Try it with two or three ants.

ANSWER: The maximum time is one minute.

6.27. HINT: Use the speed of the cyclist relative to the parade.

ANSWER: The distance traveled by the cyclist is $4 + 2\sqrt{13} \doteq 11.21$ kilometers.

6.28. HINT: The friends took twice as long to walk up the hill as to walk down the hill.

ANSWER: An arrival time of 6:30 is within half an hour of the exact arrival time. The friends walked 12 miles.

6.29. HINT: What is Bob's speed relative to the truck?
ANSWER: Bob was 0.25 kilometers behind the truck one minute before the crash.

6.30. HINT: Find relations for the downstream and upstream time in both cases.
ANSWER: The speed of the current is 2 miles per hour.

6.31. HINT: Find the net outflow rate of Tank B and the time needed to empty Tank A.
ANSWER: The time needed to empty both tanks is 45 minutes.

6.32. HINT: Find the net outflow rate of Tank B. The initial volume of water in the tanks and the certain time when Tank B has twice as much water as Tank A are unknown.
ANSWER: The water volume in Tank B is four times the water volume in Tank A when $t = \frac{600}{\pi} \doteq 54.5$ minutes.

6.33. HINT: Find Alan's and Brenda's individual harvesting rates in the garden and orchard.
ANSWER: Strategy A requires 5 hours, and strategy B requires $\frac{29}{8} \doteq 3.6$ hours.

6.34. HINT: Determine when the runners are an integer number of laps out of phase. Then determine where the passing occurs.

(i) ANSWER: The runners pass every 10 minutes at the starting point, after Polly completes 5, 10, 15, ... laps and Quinn completes, respectively, 4, 8, 12, ... laps.

(ii) ANSWER: The runners pass every minute, when Polly completes $\frac{5}{3}, \frac{10}{3}, 5, \ldots$ laps and Quinn completes, respectively, $\frac{2}{3}, \frac{4}{3}, 2, \ldots$ laps.

(iii) ANSWER: The runners pass each other every minute beginning at $\frac{1}{3}$ minutes when Polly completes

$\frac{5}{3}, \frac{20}{3}, \frac{35}{3}, \ldots$ laps and Quinn completes, respectively, $\frac{5}{3}, \frac{17}{3}, \frac{29}{3}, \ldots$ laps.

6.35. HINT: Determine the conditions that make the phases and the phase difference positive integers.

6.36. HINT: Let the track have an arbitrary length, L. The phase difference is $\phi = \theta_p + \theta_q$.
ANSWER: The runners pass each other 13 times before passing at point A.

6.37. HINT: Based on the first day's run, how much faster did Skip run than Harry?
ANSWER: The runners met four times.

6.38. HINT: The coaches take the same amount of time to reach the meeting point.
ANSWER: The distance between the towns is 168 miles.

6.39. HINT: Assign variables for the lengths and speeds of the trains, all of which are arbitrary.
ANSWER: The value of n is $1 + \sqrt{2}$.

6.40. HINT: Bret averaged 8:01 minutes for every one-mile interval, but ran at two different speeds.
ANSWER: Bret can win by one second if he runs the first 0.2 miles in 34 seconds, the next 0.8 miles in 412 seconds, and repeats the pattern for every successive mile.

6.41. HINT: One approach is to use the frame of reference of the hat.
ANSWER: He overtook his hat at 1:30 p.m.

6.42. HINT: Split the trip into three stages.
ANSWER: The trio's average speed was 9.2 miles per hour.

6.43. HINT: Label the grazing rates of the animals and the growth rate of the grass in units of pastures per day.
ANSWER: The grass can support the cow for 50 days, the horse for 100 days, and the sheep forever.

Chapter 7

At Any Rate

If a man will begin with certainties, he shall end in doubts; but if he will be content to begin with doubts, he shall end in certainties.

— Francis Bacon

The problems of the previous chapter involved constant rates of change, a feature that is often a bit of an idealization. More realistically, variables change at changing rates. Variable rate problems (specifically those that involve the motion of objects) were the original inspiration for the invention of calculus. This chapter offers many intriguing problems that involve variable rates and require the use of calculus.

Here is a fairly general principle: Suppose that $R(t)$ is the variable rate of change of a quantity Q; that is, $Q'(t) = R(t)$. Then the amount of Q that accumulates between times $t = a$ and $t = b$ is

$$\text{Amount of } Q \text{ accumulated} = \int_a^b R(t) \, dt.$$

Isaac Newton (1643–1727), living in Cambridge, England, and Gottfried Leibnitz (1646–1716), living in Paris, are generally credited with the independent and simultaneous invention (or discovery) of calculus in about 1670. However, some differ: Pierre-Simon Laplace (1749–1827) called Pierre de Fermat (1601–1665) "the true inventor of the differential calculus."

In words, the amount of Q accumulated between $t = a$ and $t = b$ is the area under the rate curve. Notice that if $R(t)$ is constant, then we have the special case studied earlier: $Q = R(b - a) = Rt$, where t is the total time elapsed. The above principle applies specifically to the case in which $R(t)$ is a velocity, in which case the accumulated amount of Q is the net distance traveled.

For example, an airplane flying into an intensifying headwind might have a decreasing speed given by $R(t) = \frac{500}{t+1}$ miles per hour. Over the course of a five-hour flight, the distance traveled (or accumulated) would be

$$\int_0^5 \frac{500}{t+1} \, dt = 500 \ln 6 \doteq 896 \text{ miles.}$$

A particularly common and important instance of variable rates arises with exponential growth and decay. Suppose that a quantity has an exponentially decreasing rate of change, $R(t) = R_0 e^{-kt}$, where $k > 0$. The amount accumulated between $t = 0$ and a future time t is

$$Q(t) = \int_0^t R(s)\, ds = \frac{R_0}{k}(1 - e^{-kt}).$$

Recall the hallmark of exponential change: A quantity that has a constant percent rate of change (for example, 5% per year) or constant relative rate of change (for example, doubles every 3 years) must be increasing or decreasing exponentially.

Notice that $Q(t)$ approaches a maximum value of $Q_{max} = \frac{R_0}{k}$ as $t \to \infty$. It can also be shown that the time at which $Q(t)$ reaches a value of $Q^* < Q_{max}$ is given by

$$t = -\frac{1}{k}\ln\left(1 - \frac{kQ^*}{R_0}\right).$$

Here are some examples.

Example 7.1 Diminishing returns. The outflow rate of a large reservoir decreases at 15% per hour, with an initial flow rate of 200 gallons per hour. After how many hours have 800 gallons drained from the reservoir? After how many hours have 1500 gallons drained from the reservoir?

Solution: The solution has two parts. The first part is to determine the outflow rate, $R(t)$; the second part is to find the amount of water released as a function of time.

The outflow rate has an exponential decrease, as implied by the words *decreases at 15% per hour*. Letting $R(t) = R_0 e^{-kt}$, with $R_0 = 200$, we must find the rate constant k. A 15% decrease per hour (every hour) implies that $R(t+1) = 0.85R(t)$ or

$$R_0 e^{-k(t+1)} = 0.85 R_0 e^{-kt}.$$

Canceling terms results in the equation $e^{-k} = 0.85$, which implies $k = -\ln 0.85 \doteq 0.1625$. Thus the outflow rate is given by

$$R(t) = 200 e^{(\ln 0.85)t}.$$

Knowing the outflow rate, we can compute the amount of water released from the reservoir. Between the initial time $t = 0$ and a future time t, the amount released is

$$Q(t) = \int_0^t R(s)\, ds = \frac{200}{\ln 0.85}(e^{\ln 0.85\, t} - 1).$$

The teacher pretended that algebra was a perfectly natural affair, to be taken for granted, whereas I didn't even know what numbers were. Mathematics classes became sheer terror and torture to me. I was so intimidated by my incomprehension that I did not dare to ask any questions.
— Carl Jung

The graphs of R and Q for these particular values of R_0 and k are shown in Figure 7.1. To find how long it takes to release 800 gallons, we must solve $Q(t) = 800$ for t. A few steps of algebra shows that

$$t = \frac{1}{\ln 0.85}\ln(1 + 4\ln 0.85) \doteq 6.461 \text{ hours.}$$

To find how long it takes to release 1500 gallons, we must solve $Q(t) = 1500$ for t. Now we find that this equation has no solution. You should verify that the maximum amount of water that can be released with this particular outflow function is $Q_{max} = -\frac{200}{\ln 0.85} \doteq 1230$ gallons.   □

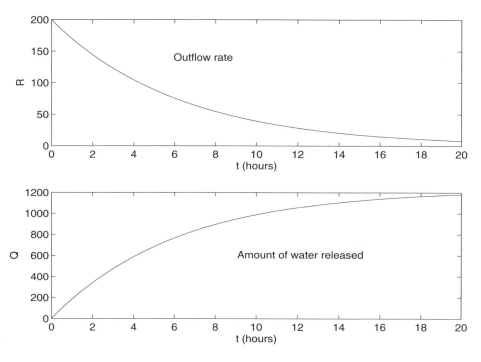

Figure 7.1. *A reservoir has an exponentially decreasing drainage rate, $R(t)$ (top). The amount of water released, $Q(t)$, approaches an asymptotic value (bottom).*

Example 7.2 Living vs. born. Which is greater, the number of people alive on the Earth in 2000 or the number of people born between 1800 and 2000? In what year did the number of people born since 1800 equal the population in that year? Use the facts that the world population was 1 billion in 1800 and 6 billion in 2000. Assume uniform exponential growth between 1800 and 2000 with a relative birth rate that is twice the relative death rate.

Solution: It seems evident that, because no one alive today was born before 1800, there are fewer people alive today than people born since 1800. However, if we choose a year closer to 1800, it is possible that the number of people alive in that year might equal the number of people born since 1800. Let's look at the mathematics.

Assuming an exponential growth law for the world population between 1800 ($t = 0$) and 2000 ($t = 200$), the population, measured in billions of people, is given by $P(t) = e^{rt}$, where the relative growth rate r must be determined. Note that $P(0) = 1$ (billion), as required. Setting $P(200) = 6$ (billion), we find that $r = \frac{\ln 6}{200} \doteq 0.009$ yr^{-1}. The growth rate represents the combined effects of births and deaths; that is, $r = b - d$, where b and d are the relative birth and death rates, respectively. Using the assumption that the birth rate is twice the death rate, we have $b - d = 0.009$ and $b = 2d$. These conditions imply that $b = 0.018$ yr^{-1} and $d = 0.009$ yr^{-1}.

To find the total number of people born between 1800 and 2000, we must identify the absolute birth rate, in units of births per year. For an exponential growth model, with $P(t) = e^{rt}$, we always have $P'(t) = re^{rt} = rP(t)$; that is, the absolute growth rate is a constant multiple of the current population. It follows that the absolute birth rate is $bP(t)$, where b is the relative birth rate determined above. Thus, the total number of people born between $t = 0$ and $t = 200$ is

$$\int_0^{200} bP(t)\, dt = \int_0^{200} be^{rt}\, dt = 10.1 \text{ billion,}$$

where the values of b and r have been used. As anticipated, the number of people born between 1800 and 2000 exceeds the world population in 2000, which was 6 billion.

To find the time t between 1800 and 2000 when the number of people born since 1800 equaled the current population, we must solve

$$\underbrace{\int_0^t bP(s)\, ds = \int_0^t be^{rs}\, ds}_{\text{number born}} = \underbrace{P(t) = e^{rt}}_{\text{current population}}.$$

Using the known values of b and r, evaluating the integral, and solving for t gives $t = 77$ years. Thus, in about 1877, the total number of people born since 1800 equaled the current population. Some interesting generalizations of this problem are given in Exercise 7.10.   □

Example 7.3 The snowplow problem. With snow on the ground and falling at a constant rate, a snowplow began plowing down a long straight road at 12:00. The plow traveled twice as far in the first hour (between 12:00 and 1:00) as it did in the second hour (between 1:00 and 2:00). At what time did the snow start falling? Assume a plowing rate that is inversely proportional to the depth of the snow.

Solution: It's easiest to let $t = 0$ be the time at which the snow started falling and $t = T$ be 12:00; the goal is to find T. Because the snow falls at a constant rate, the snow depth is given by $s(t) = at$, where a is a constant; note that $s(0) = 0$. The fact that the plowing rate is inversely proportional to the snow depth means that the speed of the plow is given by $v(t) = \frac{k}{s(t)}$, where k is a constant. Let d_1 be the distance traveled by the plow between 12:00 ($t = T$) and 1:00 ($t = T + 1$). Let d_2 be the distance traveled by the plow between 1:00 and 2:00 ($t = T + 2$). Because $d_1 = 2d_2$, we have

$$\int_T^{T+1} v(t)\, dt = 2 \int_{T+1}^{T+2} v(t)\, dt.$$

Substituting for $v(t)$ and evaluating the integrals leads to the following equation:

$$\ln\left(\frac{T+1}{T}\right) = 2\ln\left(\frac{T+2}{T+1}\right).$$

Simplifying this equation (by exponentiating both sides) results in the quadratic equation $T^2 + T - 1 = 0$. The roots are $T = \frac{1}{2}(-1 \pm \sqrt{5})$, of which only the larger root, $T = \frac{1}{2}(-1 + \sqrt{5}) \doteq 0.62$, is relevant. The snow started falling 0.62 hours before 12:00, or 11:23.   □

The previous example involved rates of change with respect to time. However, rates of change with respect to a spatial variable are equally common and important, as shown in the next example.

Example 7.4 The towed boat. Zorba is standing on a long dock holding a 20-meter rope that is attached to a boat. Zorba begins walking along the dock, towing the boat, and keeping the rope taut. At all times the path of the boat is in the direction of the rope. Describe the path taken by the boat. How far has Zorba walked when the boat is 10 meters from the dock?

Solution: A couple of assumptions and observations are needed. As shown in Figure 7.2, the dock is represented by the positive x-axis, and the boat is a point with coordinates (x, y). Note also that the distance between Zorba and the boat is always 20 meters, Zorba walks along the dock in the positive x direction, and the boat's initial position is $(0, 20)$.

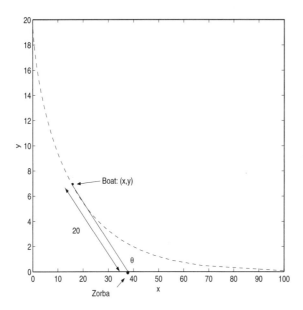

Figure 7.2. *As Zorba walks along the dock (x-axis), he pulls the boat with a taut 20-foot rope that is always tangent to the path of the boat.*

At any particular position of the boat, the tangent to the boat's path is in the direction of the rope. Thus, $\frac{dy}{dx} = \tan\theta$, where θ is the angle the rope makes with the positive x-axis. Notice that $\frac{\pi}{2} \leq \theta < \pi$, which implies that $\tan\theta < 0$, and thus we can write

$$\frac{dy}{dx} = \tan\theta = -\frac{y}{\sqrt{20^2 - y^2}} = -\frac{y}{\sqrt{400 - y^2}},$$

where the negative sign gives $\tan\theta$ the required sign. The task reduces to finding a function with the indicated derivative. It's easiest to view x as the dependent variable and y as the independent variable; we then have

$$\frac{dx}{dy} = -\frac{\sqrt{400 - y^2}}{y}.$$

The path followed by Zorba's boat is called a *tractrix*. It was first studied by Christian Huygens in 1692. Leibnitz used it to find the path of a dragged object.

Integrating with respect to y gives

$$x = -\sqrt{400 - y^2} + 20 \ \ln \frac{20 + \sqrt{400 - y^2}}{y} + C,$$

where C is a constant of integration. The condition that $x = 0$ when $y = 20$ implies that $C = 0$. The path of the boat is given by

$$x = -\sqrt{400 - y^2} + 20 \ \ln \frac{20 + \sqrt{400 - y^2}}{y},$$

as shown by the dotted line in Figure 7.2. Now let d be the distance Zorba has walked along the dock when the boat is at the point (x, y). Then

$$d = x + \sqrt{400 - y^2} = 20 \ \ln \frac{20 + \sqrt{400 - y^2}}{y}.$$

When the boat is 10 meters from the dock ($y = 10$), we find that Zorba has walked $d = 26.34$ meters. $\square$

7.1 Exercises

7.1. ◇ Population of one. Use the facts that the world population was 1 billion in 1800 and 6 billion in 2000, and assume exponential growth with a constant rate. According to this model, in what year was the world population equal to 1?

7.2. Where do they meet? Abe left the town of Arcadia, walking at 4 miles per hour towards the town of Seldom. Sally left Seldom, walking at 3 miles per hour towards Arcadia. Due to fatigue, the speed of each walker decreased to $1/(n+1)$ of its initial speed after n hours. What fraction of the distance between the two towns had Abe walked when he met Sally?

7.3. ◇ Where do they meet? Abe left the town of Felicity, walking at 4 miles per hour towards the town of Cheer. Ten miles away, Sally left Cheer, walking at 3 miles per hour towards Felicity. Due to fatigue, Abe's speed decreased by 10% each hour, while Sally's speed decreased by 19% each hour. How far from Felicity had Abe walked when he met Sally?

7.4. Shortest path. A house is located at each corner of a square with side lengths of one mile. What is the length of the shortest road system that connects all of the inhabitants? [13]

7.5. Tangency questions.

(i) ◇ It is easily verified that the graphs of $y = x^2$ and $y = e^x$ have no points of intersection, while the graphs of $y = x^3$ and $y = e^x$ have two points of intersection. It follows that for some real number $2 < p < 3$, the graphs of $y = x^p$ and $y = e^x$ have exactly one point of intersection. Using analytical and/or graphical methods, determine p and the coordinates of the single point of intersection.

(ii) It is easily verified that the graphs of $y = 1.1^x$ and $y = x$ have two points of intersection, while the graphs of $y = 2^x$ and $y = x$ have no points of intersection. It follows that for some real number $1 < p < 2$, the graphs of $y = p^x$ and $y = x$ have exactly one point of intersection. Using analytical and/or graphical methods, determine p and the coordinates of the single point of intersection.

7.6. More snowplow problems. With snow on the ground and falling at a constant rate, a snowplow began plowing down a long straight road at 12:00. The plow traveled twice as far in the first hour (between 12:00 and 1:00) as it did in the second hour (between 1:00 and 2:00). At what time did the snow start falling? Assume a plowing rate that is inversely proportional to the *square* of the snow depth.

7.7. ◇ Depletion of natural resources. Suppose that the rate, $r(t)$, at which a nation extracts oil declines exponentially as $r(t) = r_0 e^{-kt}$, where $r_0 = 10^7$ tons per year is the current rate of extraction. Suppose also that the estimate of the total oil reserve is 2×10^9 tons.

(i) Find the maximum decay constant k for which the total oil reserves will last forever.

(ii) Suppose that the rate of extraction declines at a rate that is twice the value $(2r(t))$ in part (i). How long will the total oil reserves last?

7.8. Slowest shortcut. Weary Willie, sleeping $\frac{1}{4}$ mile from the tracks, was awakened just as the locomotive of a train was at the point on the tracks nearest him. The train was $\frac{1}{3}$ mile long and traveled 20 miles per hour. If he started running immediately and cut across a field in a straight line, how slowly could he run and still catch the train? In what direction should he run?

7.9. ◇ Slicing a pizza. A 12-inch (diameter) pizza is cut into three pieces by making two parallel cuts, d inches to the right and left of the center of the pizza.

(i) Which of the three pieces is largest if $d = 1.8$ inches? [7]

(ii) Suppose that 1 inch of crust is 1.5 times as valuable to you (in terms of eating pleasure) as 1 square inch of pizza. Which of the pieces is most valuable if $d = 1.8$ inches?

7.10. Born vs. living questions. Suppose that a population is described by an exponential growth law $p(t) = p_0 e^{rt}$, with $r = b - d$, where b and d are the relative birth and death rates, respectively.

(i) Show that the time $t^* > 0$ at which the number of people born since $t = 0$ equals the current population is given by $t^* = \frac{1}{r} \ln\left(\frac{b}{d}\right)$, which requires that $b > d$ for this situation to occur.

(ii) Let t^* be the time at which the number of people born since $t = 0$ equals the current population, $p(t^*)$. Show that $p(t^*) = (b/d)p_0$.

7.11. ◇ Optimal can.

(i) **Original problem.** Find the radius and height of a cylindrical soda can with a volume of 354 cubic centimeters that minimize the surface area.

(ii) **Real problem.** Compare your answer in part (i) to a real soda can, which has a volume of 354 cubic centimeters, a radius of 3.1 centimeters, and a height of 12.0 centimeters, to conclude that real soda cans do not *seem* to have an optimal design. Now account for the fact that real soda cans have a double thickness in their top and bottom surfaces. Are the dimensions of the real can closer to optimal? [suggested by Bruce MacMillan, University of Colorado at Denver]

7.12. Turning a boat. A large boat 150 feet in length changes its direction by 30 degrees while moving a distance of twice its length. What is the radius of the circle along which it moves? [14]

7.13. ◇ Turning a corner. What is the length of the longest pole that can be carried *horizontally* around a corner where a four-foot-wide corridor and a three-foot-wide corridor meet at right angles?

7.14. The grazing goat. A goat is tied by a rope to the inside perimeter of a circular corral of unit radius.

(i) What is the area of the region within the corral grazed by the goat?

(ii) How long should the rope be in order to insure that the goat grazes exactly $\pi/4$ square units?

7.15. ◇ Covering a marble. Imagine a flat-bottomed pot with a circular cross-section of radius 4 inches. What is the radius of the marble, with radius $0 < r \le 4$ inches, which when placed in the bottom of the pot, requires the largest amount of water to be completely covered?

7.16. Circle and square. A piece of wire 60 centimeters in length must be cut, with the resulting two pieces formed into a circle and a square. Where should the wire be cut to (i) minimize and (ii) maximize the combined area of the circle and the square?

7.17. ◇ Walking and rowing. A boat is 4 miles from the nearest point on a straight shoreline; that point is 6 miles from a restaurant on the shore. A woman plans to row the boat straight to a point on the shore, and then walk along the shore to the restaurant. If she can walk at 3 miles per hour, what is the minimum speed at which she must row so that the quickest way to the restaurant is to row directly (with no walking)? [7]

7.18. Walking and swimming. A man wishes to get from an initial point on the shore of a circular pond with a radius of 1 mile to a point on the shore directly opposite (on the other end of the diameter). He plans to swim from the initial point to another point on the shore and then walk along the shore to the terminal point.

(i) If he swims at 2 miles per hour and walks at 4 miles per hour, what are the minimum and maximum times for the trip? [7]

(ii) If he swims at 2 miles per hour and walks at $\frac{3}{2}$ miles per hour, what are the minimum and maximum times for the trip?

(iii) If he swims at 2 miles per hour, what is the minimum walking speed for which it is quickest to walk the entire distance?

7.19. ◇ Folded boxes.

(i) **The original problem.** From a rectangular piece of cardboard measuring 3 feet by 4 feet, a square with sides of length x is cut out of each corner. The resulting piece of cardboard is then folded into a box without a lid. Find the largest box volume that can be formed in this way.

(ii) **Start with a square.** Suppose that in part (i) the original piece of cardboard is a square with sides of length a. Find the volume of the largest box that be formed in this way.

(iii) **In the limit.** Suppose that in part (i) the original piece of cardboard is a rectangle with sides of length a and b. Holding a fixed, find the size x of the corner squares that maximizes the volume of the box as $b \to \infty$. [*Mathematics Teacher*, November 2002 p. 568]

7.20. Golden earring. A circle of radius r is removed from a larger circle of radius R as shown in the figure below. Find the ratio of the radii, $\frac{R}{r}$, such that the center of gravity (balancing point) of the resulting earring is located at the point P. [P. Glaister, "Golden Earrings," *Mathematical Gazette*, 80, 1996, pp. 224–225]

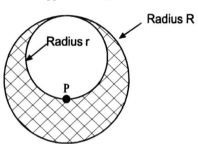

7.21. ◇ An ant's walk. If an ant lands randomly within a circle of radius R and crawls in a straight line in a random direction, what is the average (expected) distance it travels to the edge of the circle? [12]

7.22. Pursuit problem. In some pursuit problems, the goal of the pursuer is to catch the pursued (see Chapters 9 and 12 for examples). In other pursuit problems, such as this one, the goal is to keep the pursued in sight. Suppose that a photographer is 100 meters due north of a large elk, which is walking due west. The photographer walks directly toward the elk at all times, maintaining a constant distance of 100 meters.

(i) Describe the path of the photographer.

(ii) What is the photographer's position when she is walking in a direction 10 degrees south of due west?

7.2 Hints and Answers

7.1. HINT: Find the rate constant for the exponential growth law.
ANSWER: The population was 1 person in 513 B.C.E.
7.2. HINT: Abe's speed was $v_a(t) = \frac{4}{t+1}$ and Sally's speed was $v_s(t) = \frac{3}{t+1}$.
ANSWER: Abe traveled $\frac{4}{7}$ of the distance between the towns.
7.3. HINT: Abe's speed was $v_a(t) = 4e^{t\ln 0.9} = 4(0.9)^t$ and Sally's speed was $v_s(t) = 3e^{t\ln 0.81} = 3(0.81)^t$.
ANSWER: Abe and Sally met approximately 5.912 kilometers from Felicity.
7.4. HINT: Use symmetry and draw diagonal roads from the towns to a central road.

ANSWER: The length of the shortest road system is $1 + \frac{3}{\sqrt{3}}$ miles.
7.5. (i) HINT: Write conditions for the intersection and tangency of the curves $y = e^x$ and $y = x^p$.
ANSWER: The curves $y = e^x$ and $y = x^e$ are tangent at the point (e, e^e).

(ii) HINT: Write conditions for intersection and tangency of the curves $y = p^x$ and $y = x$.
ANSWER: With $p = e^{1/e}$, the curves $y = p^x$ and $y = x$ are tangent at the point (e, e).

7.6. HINT: Let $t = 0$ be the time at which the snow started falling. Let $t = T$ be noon, and find T.

ANSWER: The snow started falling at 10:00.

7.7. HINT: Integrate the rate of extraction to find the total amount of oil extracted.

ANSWER: (i) $k = 0.005$ yr^{-1}, (ii) 6.0 years.

7.8. HINT: Find the position of the end of the train, which is where Willie should catch the train.

ANSWER: Willie can run no slower than 12 miles per hour, and he must run toward a point 0.2 miles beyond the point on the tracks nearest to where Willie slept.

7.9. HINT: Use symmetry. The area of a sector of a circle of radius r subtended by an angle θ is $A = \frac{1}{2}r^2\theta$. The length of an arc of a circle of radius r subtended by an angle θ is $s = r\theta$.

ANSWER: (i) The center region is larger. (ii) The left and right regions (equal in value) are more valuable.

7.10. HINT: Use the absolute birth rate, which is $bp(t)$, where $p(t)$ is the population.

7.11. HINT: Minimize the surface area subject to the volume constraint.

ANSWER: (i) $r = 3.83$ centimeters, $h = 7.67$ centimeters; (ii) $r = 3.04$ centimeters, $h = 12.1$ centimeters.

7.12. HINT: The length of an arc of a circle of radius r subtended by an angle θ is $s = r\theta$.

ANSWER: The radius of the circle along which the boat moves is 573 feet.

7.13. HINT: For a given angle at which the pole enters the hallway, find the *maximum* length of a pole that turns the corner.

ANSWER: The pole of maximum length that turns the corner is 9.87 feet long.

7.14. HINT: The problem can be done with or without calculus.

ANSWER: (i) The area is 1.23 square units. (ii) The radius should be 0.78 units.

7.15. HINT: Find the volume of water in the pot when the marble is just covered with water.

ANSWER: A marble of radius 2.82 inches requires the most water to cover it.

7.16. HINT: Find the total area as a function of x, the amount of wire used for the square.

ANSWER: (i) To minimize the total area, use $\frac{240}{4+\pi} \doteq 33.6$ centimeters for the square. (ii) To maximize the total area, use all of the wire for the circle.

7.17. HINT: Find the total travel time, and minimize it with respect to the landing point on the shore.

ANSWER: She must row at least $\frac{9}{\sqrt{13}}$ miles per hour.

7.18. HINT: Find the swimming distance as a chord length and the walking distance as an arc length, both as functions of a central angle of the circle.

(i) ANSWER: The minimum travel time is $\frac{\pi}{4}$ hours (achieved by walking around the edge of the lake). The maximum travel time is $\frac{\sqrt{3}}{2} + \frac{\pi}{12} \doteq 1.12$ hours.

(ii) ANSWER: The minimum travel time is one hour (swimming the entire way) and the maximum travel time is $\frac{2\pi}{3} \doteq 2.1$ hours (walking the entire way).

(iii) ANSWER: He must walk at least 2 miles per hour.

7.19. HINT: Find the volume of the box as a function of x, the size of the corner squares.

ANSWER: (i) The largest box has a volume of approximately 3.03 cubic feet. (ii) The largest box has a volume of $\frac{2a^3}{27}$ cubic feet. (iii) The size of the corner squares that maximizes the volume of the box in this limit is $\frac{a}{4}$.

7.20. HINT: Place the origin of an (x, y) coordinate system at the center of the large circle. The earring is symmetric about the y-axis. The position of the center of mass in the y-direction is $y^* = \frac{1}{A} \iint_D y \, dA$, where D is the region formed by the earring and A is its area. Use polar coordinates.

ANSWER: The required ratio of the radii is $\frac{r}{R} = \frac{1}{2}(1 + \sqrt{5})$, the golden mean!

7.21. HINT: The average value of a function f over an interval (a, b) is $\frac{1}{b-a} \int_a^b f(x) \, dx$.

ANSWER: The average distance traveled is $\frac{8R}{3\pi}$.

7.22. HINT: If the path is given by $y = f(x)$, determine $y'(x)$ in terms of y.

ANSWER: (i) The path is given by

$$x = \sqrt{10{,}000 - y^2} - 100 \ln \frac{100 + \sqrt{10{,}000 - y^2}}{y}.$$

(ii) To the nearest integer, the photographer's position is $(-430, 1)$.

Chapter 8

Difference Equations

I have deeply regretted that I did not proceed far enough at least to understand something of the great leading principles of mathematics; for [those] thus endowed seem to have an extra sense.

— Charles Darwin

8.1 Background

In this chapter, we explore a powerful problem-solving and modeling tool known as **difference equations**. As you will see, difference equations arise in problems that range from drug metabolism to house loans. They also allow us to work with some truly elegant mathematics.

Difference equations are also called recurrence relations, iterations, and recursion formulas. The choice of terminology often depends on the context of the problem.

A difference equation is a rule that defines a sequence of numbers. For example, suppose you put \$100 in a bank account that earns interest at an annual rate of 5% (compounded annually). If we let B_n be the balance in the account after n years, then we can write

$$B_0 = \$100 \quad \text{(initial balance)}, \tag{8.1}$$

$$B_{n+1} = 1.05B_n \quad \text{for } n = 1, 2, 3, \ldots . \tag{8.2}$$

The first equation is called an **initial condition** because it gives the value of the initial term of the sequence. The second equation is a difference equation; it tells us how to find next year's balance in terms of this year's balance. (In this case, the new balance is found by increasing the current balance by 5%.)

Our goal in this chapter is to formulate and solve difference equations. However, we must be clear about what we mean by *solving* a difference equation. It is always possible to enumerate the solution by starting with B_0 and computing $\{B_1, B_2, B_3, \ldots\}$ using the rule provided by the difference equation. Such a process might be called **solution by enumeration**.

However, suppose that you wanted to compute, say, B_{23} or B_{1001} directly, without enumeration. To do this, it is necessary to have a formula, $B_n = F(n)$, which gives B_n for any value of n. If we can accomplish this task, we have found an **analytical** (or *closed form* or *explicit*) solution to the difference equation. Of course, once an analytical solution is found, it is always possible to check it by enumeration.

Let's begin with some definitions. The general form of a difference equation is

$$x_{n+1} = f(x_n, x_{n-1}, \ldots, x_{n-k+1}) \quad \text{for } n = k - 1, k, k + 1, \ldots.$$

Notice that the next term in the sequence, x_{n+1}, depends on the k previous terms, so this is called a **kth-order** difference equation. If the function f is linear in all of its arguments, the difference equation is **linear**; otherwise, it is **nonlinear**. For example, $x_{n+1} = x_n(2 - x_n)$ is a first-order nonlinear difference equation, while $x_{n+1} = x_n + x_{n-1}$ is a second-order linear difference equation. We will see that linear difference equations can usually be solved analytically, while nonlinear difference equations tend to be impossible to solve analytically (and are much more interesting).

Finally, difference equations are often used to model a "system" that evolves in time. In such cases, the index n represents time (perhaps in hours, years, or generations), and the solution x_n represents the state of the system, perhaps the population of a community or the balance in a bank account. For this reason, x_n is often called the **state variable**. As mentioned earlier, the first term of the solution, x_0, represents the initial state of the system and is called the initial condition. Let's proceed by example.

My specific goal is to revolutionize the future of the species. Mathematics is just another way of predicting the future.
— Ralph Abraham

Example 8.1 Exponential growth and the CPI. Many variables increase by a fixed percentage every fixed period of time. For example, consider the Consumer Price Index (CPI), which is often used as a measure of the cost of living and inflation. Computed monthly by the Bureau of Labor Statistics using the prices of over 60,000 items, the index uses the years 1983–1984 as a baseline with a value of 100 (see Figure 8.1). A rough calculation indicates that the annual growth rate of the CPI has averaged about 3.2% since 1983. In fact, the smooth curve in Figure 8.1 is an exponential function that increases by 3.2% per year. It gives a reasonably good fit to the data. Find and solve the difference equation for the CPI.

Solution: We let x_n denote the CPI for n years after 1983, for $n = 0, 1, 2, \ldots$, with $x_0 = 100$. Then the difference equation that models the growth of the CPI is

$$x_{n+1} = x_n + 0.032x_n = 1.032x_n \quad \text{for } n = 0, 1, 2, \ldots;$$

that is, the CPI for the next year is the current CPI plus 3.2% of the current CPI.

If we list a few terms of the sequence, the general pattern will become evident. Assuming that $x_0 = 100$, we have

$$x_1 = 1.032x_0,$$
$$x_2 = 1.032x_1 = 1.032^2 x_0,$$
$$x_3 = 1.032x_2 = 1.032^3 x_0,$$
$$\vdots \quad \vdots \quad \vdots$$
$$x_n = 1.032x_{n-1} = 1.032^n x_0.$$

From the last line, we can identify the analytical solution to this difference equation; it is $x_n = 1.032^n x_0$. This solution accomplishes our goal: It enables us to compute x_n

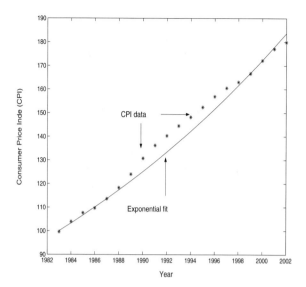

Figure 8.1. *The Consumer Price Index (CPI) is shown (∗) for the years* 1983 *through* 2002. *The smooth curve gives an exponential fit to the data, assuming an average annual growth rate of 3.2%.*

directly, without enumeration, for any value of n. Written in this form, the sequence $\{x_n\}$ clearly exhibits exponential growth; it increases by a fixed percentage with each term, and it has a constant doubling time.   □

More generally, consider a quantity that increases by $p\%$ per time period. Letting $a = p/100$, the resulting difference equation is

$$x_{n+1} = (1 + a)x_n \quad \text{for } n = 0, 1, 2, \ldots,$$

which has the solution $x_n = (1+a)^n x_0$. The resulting sequence exhibits exponential growth.

Similarly, if a quantity decreases by $p\%$ per time period, and we let $a = p/100$, then the governing difference equation is

$$x_{n+1} = (1 - a)x_n \quad \text{for } n = 0, 1, 2, \ldots.$$

The solution is $x_n = (1 - a)^n x_0$, and the state variable decays exponentially.

Example 8.2 Supply and depletion problems. Models of many practical problems rely on the difference equation

$$x_{n+1} = ax_n \pm b \quad \text{for } n = 0, 1, 2, \ldots,$$

where $a > 0$ and $b > 0$ are given constants. The choice of $\pm$ in the difference equation allows many different situations to be modeled. Here are three examples:

- Consider paying into a periodic savings plan at a rate of $50 per month, where the account earns 1% interest per month. Letting x_n be the balance in the account after the nth month, it increases according to the difference equation

$$x_{n+1} = 1.01x_n + 50 \quad \text{for } n = 0, 1, 2, \ldots, \text{ with } x_0 = 0.$$

  The equation says that the new balance is found by increasing the current balance by 1% and adding $50.

- Suppose you are paying off an $80,000 loan with monthly payments of $900 and a monthly interest rate of 0.67% (8% annual interest rate). Letting x_n be the balance in the account after the nth month, the loan balances are governed by the difference equation

$$x_{n+1} = 1.0067x_n - 900 \quad \text{for } n = 0, 1, 2, \ldots, \text{ with } x_0 = \$80,000.$$

  The equation says that the new loan balance is found by increasing the current balance by 0.67% and subtracting $900.

You cannot apply mathematics as long as words still becloud reality.
— Hermann Weyl

- Suppose you take a 200-milligram dose of an antibiotic every 12 hours, and that the amount of drug in the blood decreases to 40% of its current level every 12 hours. Letting x_n be the amount of antibiotic in your blood after the nth dose, the difference equation for this system is

$$x_{n+1} = 0.40x_n + 200 \quad \text{for } n = 0, 1, 2, \ldots, \text{ with } x_0 = 0.$$

  The equation says that the new drug amount is 40% of its current value plus 200 milligrams.

In each case, a represents a natural growth or decay rate of the system, while b represents a supply to or depletion of the system. Having seen just three of the many diverse applications of this difference equation, how is it solved analytically?

Solution: Let's write a few terms of the sequence generated by the difference equation

$$x_{n+1} = ax_n + b \quad \text{for } n = 0, 1, 2, \ldots.$$

Assuming x_0 is given, we have

$$
\begin{aligned}
x_1 &= ax_0 + b, \\
x_2 &= ax_1 + b = a\underbrace{(ax_0 + b)}_{x_1} + b = a^2 x_0 + b(1 + a), \\
x_3 &= ax_2 + b = a\underbrace{(a^2 x_0 + b(1 + a))}_{x_2} + b = a^3 x_0 + b(1 + a + a^2), \\
&\vdots \quad \vdots \quad \vdots \\
x_n &= ax_{n-1} + b = a^n x_0 + b(1 + a + a^2 + \cdots + a^{n-1}).
\end{aligned}
$$

At each step we have substituted the solution from the previous step, until a pattern emerges. The last line gives a formula for x_n in terms of only n and x_0, so it is an analytical solution. However, we can clean it up a bit by recalling how to evaluate a geometric sum. Knowing that

$$1 + a + a^2 + \cdots + a^{n-1} = \frac{a^n - 1}{a - 1},$$

we can write the solution to the difference equation as

$$x_n = x_0 a^n + b\frac{a^n - 1}{a - 1} \quad \text{for } n = 0, 1, 2, \ldots.$$

A similar analysis with the equation $x_{n+1} = ax_n - b$ leads to the solution

$$x_n = x_0 a^n - b\frac{a^n - 1}{a - 1} \quad \text{for } n = 0, 1, 2, \ldots.$$

There is a useful interpretation of both of these solutions: The term involving x_0 represents the effect of the initial condition; we see that it is amplified or diminished by a factor of a at each step. The second term, involving b, represents the accumulated supply or depletion over n steps of the process. Let's consider the three examples given above.

- For the savings plan problem with $a = 1.01$, $b = 50$, and $x_0 = 0$, the general solution is
$$x_n = 50\frac{1.01^n - 1}{0.01} \quad \text{for } n = 0, 1, 2, \ldots.$$
Plotted in Figure 8.2 (a), the solution consists of discrete points. We see that after 240 months (20 years), the balance in the account reaches about $50,000.

- For the loan problem, with an initial loan amount of $x_0 = \$80,000$, $a = 1.0067$, and $b = 900$, the general solution is
$$x_n = 80{,}000(1.0067^n) - 900\frac{1.0067^n - 1}{0.0067} \quad \text{for } n = 0, 1, 2, \ldots.$$
The solution is plotted in Figure 8.2 (b). We see that after about 137 months, the loan balance reaches zero and the loan is paid off.

- For the periodic dosing problem, with no antibiotic initially in the blood ($x_0 = 0$), $a = 0.4$, and $b = 200$ milligrams, the general solution is
$$x_n = 200\frac{0.4^n - 1}{0.60} \quad \text{for } n = 0, 1, 2, \ldots.$$
The solution is plotted in Figure 8.2 (c). The amount of antibiotic in the blood levels off at a steady state value of $\frac{100}{0.6} = 167$ milligrams. □

Having discussed first-order difference equations, let's move on to some second-order linear problems, which have the general form

$$x_{n+1} = ax_n + bx_{n-1} \quad \text{for } n = 1, 2, 3 \ldots.$$

Generating a solution by enumeration requires *two* initial conditions, x_0 and x_1, to start the process. In general, we should expect to supply k initial conditions for a kth-order problem. We will investigate the solution of second-order problems with perhaps the oldest and most famous difference equation. It was proposed in 1202 A.D. by Fibonacci. Here is the problem he posed.

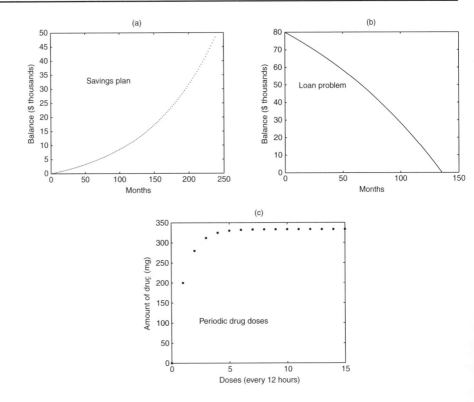

Figure 8.2. *Solutions to difference equations that model a savings plan* (a), *a loan payoff* (b), *and periodic doses of a drug* (c). *In each case, the solution consists of discrete points.*

Leonardo Pisano, better known as Fibonacci, was born in about 1170 in Italy but spent his early years in North Africa. In 1200 he returned to Pisa, where he wrote several influential mathematics books. His *Liber abaci* introduced to Europe the Hindu-Arabic place-valued decimal system and the use of Arabic numerals; it also posed the famous rabbit problem. Fibonacci died in about 1250 in Pisa.

For those who are fanatic about Fibonacci numbers, there is an entire journal, the *Fibonacci Quarterly*, devoted to them.

Example 8.3 Fibonacci's rabbits. Suppose that two newly born rabbits, one male, one female, are put in a field. Beginning in their second month, a pair of rabbits can produce another pair of rabbits (a male and a female) every month thereafter. Assuming that rabbits never die, how many pairs of rabbits will there be after n months?

Solution: If you draw some pictures and apply the rules for reproduction, you will see that the monthly rabbit population is

$$1, \ 1, \ 2, \ 3, \ 5, \ 8, \ 13, \ 21, \ldots.$$

This solution is the famous **Fibonacci sequence**, and it can be generated by the difference equation

$$x_{n+1} = x_n + x_{n-1} \quad \text{for } n = 1, 2, 3 \ldots,$$

with $x_0 = x_1 = 1$. It's easy to enumerate the solution, but what if you are asked to find, say, x_{1001} directly, without enumeration? To do this, an analytical solution is needed. Let's see how it can be found.

You may be familiar with a strategy called *trial solutions*. We take an (educated) guess about the possible form of the solution, substitute it into the equation, and see where it leads. For linear difference equations, a good trial solution is $x_n = p^n$ (an exponential function), where the constant p needs to be determined. Substituting the trial solution into the difference equation gives us

$$p^{n+1} = p^n + p^{n-1}.$$

Assuming $p \neq 0$ and canceling the common term p^{n-1} results in a quadratic polynomial in p:

$$p^2 - p - 1 = 0.$$

This polynomial is the **characteristic polynomial** for the problem, and its roots give us two values for p:

$$\{p_+, p_-\} = \frac{1 \pm \sqrt{5}}{2}.$$

The positive root is called the **golden mean** and is closely associated with the Fibonacci sequence.

We now have solutions of the form $x_n = p_+^n$ and $x_n = p_-^n$. Because the problem is linear, it is not difficult to show that $x_n = ap_+^n$ and $x_n = bp_-^n$ are also solutions, where a and b are arbitrary real numbers. In fact, we can combine the two solutions and write the **general solution** for the problem; it is

$$x_n = ap_+^n + bp_-^n.$$

The arbitrary constants, a and b, can be determined using the initial conditions. The condition $x_0 = 1$ implies that $x_0 = a+b = 1$, while the condition $x_1 = 1$ implies that $ap_+ + bp_- = 1$. Thus, we have two equations to solve for a and b. A little bit of algebra reveals that

$$a = \frac{1}{2}\left(1 + \frac{1}{\sqrt{5}}\right) \quad \text{and} \quad b = \frac{1}{2}\left(1 - \frac{1}{\sqrt{5}}\right).$$

With a and b determined, we have the solution

$$x_n = \left(\frac{1 + \frac{1}{\sqrt{5}}}{2}\right)\left(\frac{1 + \sqrt{5}}{2}\right)^n + \left(\frac{1 - \frac{1}{\sqrt{5}}}{2}\right)\left(\frac{1 - \sqrt{5}}{2}\right)^n.$$

This is a remarkable result: Despite the powers of irrational quantities in this expression, it generates nothing but integers!   □

> The Greeks claimed that the most visually pleasing division of a line segment has the property that the ratio of the length of the long piece to the length of the short piece is the same as the ratio of the length of the whole segment to the length of the long piece. This division results in the golden mean.

Any linear difference equation with constant coefficients (independent of n) can be solved using the method of trial solutions. A kth-order linear difference equation produces a kth-degree characteristic polynomial, and the general solution has k independent terms. An extensive theory exists for dealing with cases in which the characteristic polynomial has repeated real or complex roots.

There are two ways to proceed if we are looking for more complex difference equations. The first way is to consider first-order *nonlinear* difference equations, which takes us into a realm where analytical solutions are very scarce. The second way is to consider systems of first-order *linear* difference equations. This path leads to some very elegant analytical work. The following example illustrates the second type of problem.

> Geometry has two great treasures: one is the theorem of Pythagoras; the other, the division of a line into extreme and mean ratio. The first we may compare to gold; the second we may name a precious jewel.
> — Johannes Kepler

Example 8.4 Market analysis. Suppose you have been hired to do a market analysis of two brands of breakfast cereal, Granola and Wheaties. A survey that you conduct reveals the following conclusions: During any one-month period, $\frac{8}{10}$ of the Granola eaters remain Granola eaters, while $\frac{2}{10}$ of the Granola eaters switch to Wheaties. On the other hand, during any one-month period, $\frac{6}{10}$ of the Wheaties eaters remain Wheaties eaters, while $\frac{4}{10}$ of the Wheaties eaters switch to Granola. You also know that no one surveyed stops eating cereal or switches to a third brand, and that the current market share is 9 Wheaties eaters for every 1 Granola eater. What can you say about the long-term market share for each brand of cereal? Will one brand take over the entire market? Will there always be more Wheaties eaters than Granola eaters?

Solution: We can answer these questions using difference equations. We let g_n and w_n be the fraction of the market comprised of Granola eaters and Wheaties eaters, respectively, after n months. Notice that $0 \leq g_n \leq 1, 0 \leq w_n \leq 1$, and $g_n + w_n = 1$ for all n. We are given the initial conditions $g_0 = \frac{1}{10}$ and $w_0 = \frac{9}{10}$.

We now must describe the market shares for the next month in terms of the market shares for the current month. During each month, $\frac{8}{10}$ of the current Granola eaters remain with Granola, and $\frac{4}{10}$ of the current Wheaties eaters switch to Granola; this means that

$$g_{n+1} = \frac{8}{10}g_n + \frac{4}{10}w_n.$$

Similarly, during each month, $\frac{6}{10}$ of the current Wheaties eaters remain with Wheaties, and $\frac{2}{10}$ of the current Granola eaters switch to Wheaties; this means that

$$w_{n+1} = \frac{6}{10}w_n + \frac{2}{10}g_n.$$

Notice that if we add these two equations, we have $g_{n+1} + w_{n+1} = g_n + w_n$, as must be the case. We can write these equations more compactly in matrix form:

$$\begin{pmatrix} g_{n+1} \\ w_{n+1} \end{pmatrix} = \begin{pmatrix} \frac{8}{10} & \frac{4}{10} \\ \frac{2}{10} & \frac{6}{10} \end{pmatrix} \begin{pmatrix} g_n \\ w_n \end{pmatrix} \quad \text{for} \quad n = 0, 1, 2, \ldots. \tag{8.3}$$

The model that results from these two difference equations is often called a **Markov chain.**

Andrei Markov (1856–1922) graduated from Saint Petersburg University, where he later became a professor. He made wide-ranging contributions to number theory, continued fractions, and probability.

It is useful to write the equations in the vector form $\mathbf{x}_{n+1} = A\mathbf{x}_n$, where A is the 2×2 matrix of transition rates and $\mathbf{x}_n = (g_n, w_n)^T$ is the column vector of market shares after the nth month. Following the trial solution strategy used earlier, we look for a solution of the form $\mathbf{x}_n = \mathbf{v}\lambda^n$, where $\mathbf{v}$ is a column vector to be determined and λ is a scalar to be determined. Substituting this trial solution into the equation $\mathbf{x}_{n+1} = A\mathbf{x}_n$, we have $\mathbf{v}\lambda^{n+1} = A\mathbf{v}\lambda^n$. Assuming that $\lambda \neq 0$ and canceling a factor of λ^n leaves us with the equation

$$A\mathbf{v} = \lambda\mathbf{v}.$$

This important problem of linear algebra is called the **eigenvalue problem.**

Recall that the goal is to find the scalar λ and the nonzero vector $\mathbf{v}$. We can rewrite $A\mathbf{v} = \lambda\mathbf{v}$ as $(A - \lambda I)\mathbf{v} = 0$, which has nontrivial solutions, provided that $\det(A - \lambda I) = 0$. This condition is called the **characteristic polynomial** for A.

If this determinant is expanded, it becomes a second-degree polynomial in λ, whose roots are the **eigenvalues**. For each value of λ, the corresponding **eigenvectors**, $\mathbf{v}$, can be found directly. You should verify that the eigenvalues and eigenvectors associated with the matrix in (8.3) are

$$\lambda_1 = 1 \quad \text{with} \quad \mathbf{v}_1 = (2,\, 1)^T \quad \text{(or any multiple)},$$
$$\lambda_2 = 0.4 \quad \text{with} \quad \mathbf{v}_2 = (1,\, -1)^T \quad \text{(or any multiple)}.$$

Thus, the general solution of the difference equation is

$$\mathbf{x}_n = a\mathbf{v}_1\lambda_1^n + b\mathbf{v}_2\lambda_2^n = a \begin{pmatrix} 2 \\ 1 \end{pmatrix} + b \begin{pmatrix} 1 \\ -1 \end{pmatrix} 0.4^n,$$

where a and b are arbitrary constants, and we have used the fact that $\lambda_1^n = 1$. The constants a and b can be found using the initial conditions:

$$\mathbf{x}_0 = \begin{pmatrix} 0.1 \\ 0.9 \end{pmatrix} = a \begin{pmatrix} 2 \\ 1 \end{pmatrix} + b \begin{pmatrix} 1 \\ -1 \end{pmatrix}.$$

Solving these two equations gives us $a = \frac{1}{3}$ and $b = -\frac{17}{30}$. Thus, the solution to the difference equation with the given initial conditions is

$$\mathbf{x}_n = \frac{1}{3} \begin{pmatrix} 2 \\ 1 \end{pmatrix} - \frac{17}{30} \begin{pmatrix} 1 \\ -1 \end{pmatrix} 0.4^n.$$

With this solution, we can determine the market shares for every month. Moreover, we see that in the long run the market shares approach

$$\lim_{n \to \infty} \mathbf{x}_n = \frac{1}{3} \begin{pmatrix} 2 \\ 1 \end{pmatrix} - \frac{17}{30} \begin{pmatrix} 1 \\ -1 \end{pmatrix} \lim_{n \to \infty} 0.4^n$$
$$= \frac{1}{3} \begin{pmatrix} 2 \\ 1 \end{pmatrix} = \begin{pmatrix} 2/3 \\ 1/3 \end{pmatrix}.$$

Over time, the market approaches a state in which two-thirds of consumers buy Granola and one-third buy Wheaties. The solution to the problem is shown in Figure 8.3, where the pairs (g_n, w_n) are plotted for each month. We see that as time progresses, the market reaches a steady state of $\left(\frac{2}{3}, \frac{1}{3}\right)$.

There are several properties that are common to all Markov models. Notice that we could have found the steady state solution without actually solving the difference equations. If we return to the system (8.3) and let $n \to \infty$, we have

$$\lim_{n \to \infty} \begin{pmatrix} g_{n+1} \\ w_{n+1} \end{pmatrix} = \lim_{n \to \infty} \begin{pmatrix} \frac{8}{10} & \frac{4}{10} \\ \frac{2}{10} & \frac{6}{10} \end{pmatrix} \begin{pmatrix} g_n \\ w_n \end{pmatrix}.$$

Letting $\lim_{n \to \infty} g_n = g_\infty$ and $\lim_{n \to \infty} w_n = w_\infty$ results in the system of linear equations

$$\begin{pmatrix} g_\infty \\ w_\infty \end{pmatrix} = \begin{pmatrix} \frac{8}{10} & \frac{4}{10} \\ \frac{2}{10} & \frac{6}{10} \end{pmatrix} \begin{pmatrix} g_\infty \\ w_\infty \end{pmatrix}.$$

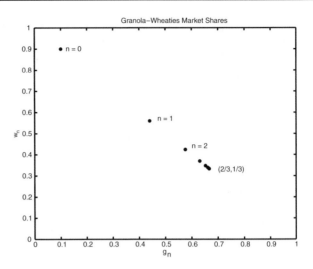

Figure 8.3. *The solution to the coupled difference equations modeling market shares of breakfast cereal is a sequence of points converging to $\left(\frac{2}{3}, \frac{1}{3}\right)$.*

The solution to this system is found to be $g_\infty = \frac{2}{3}$ and $w_\infty = \frac{1}{3}$.

More generally, the matrix A of transition rates in a Markov model must have the property that the column sums are 1 (why?); thus A must have the form

$$A = \begin{pmatrix} \alpha & \beta \\ 1 - \alpha & 1 - \beta \end{pmatrix}.$$

It is easily shown that, for matrices of this form, $\lambda_1 = 1$ and $0 < \lambda_2 < 1$. Thus, as $n \to \infty$, the solution approaches a multiple of the eigenvector $\mathbf{v}_1$, which gives us another way to determine the ratio of the market shares in the long run.   □

8.2   Exercises

8.1. ◇ Another rabbit problem. Each year a colony of rabbits grows in number by 5% of its current population due to births but is also depleted by 3.5% of its current population due to deaths. If the first observed population of the colony is 650 rabbits, what is the population of the colony n years later? When will the population reach 2500?

8.2. Fish harvesting. A fishery manager knows that her fish population naturally increases at a rate of 1.5% per month. At the same time, fish are harvested at a rate of 80 fish per month. Suppose that the initial fish population is 4000 fish and that these rates remain constant.

(i) Find an expression for the fish population after the nth month, for any $n = 0, 1, 2, \ldots$.

(ii) What is the fish population after 23 months (to the nearest whole fish) and in the steady state?

(iii) What is the maximum harvesting rate that allows the population to increase?

8.3. ◇ Savings plan. Suppose that each month you deposit $100 into an account that earns interest at an *annual* rate of 9.0%.

(i) Find the balance in the account after n months for any positive integer n.

(ii) How long must you make deposits to reach a balance of $20,000?

8.4. House loan. A house loan for $120,000 must be paid back in monthly payments of $800 at an annual interest rate of 7.5%. How many months are needed to retire the loan?

8.5. ◇ Equal loan payments. Jon and Linda take out a home mortgage for $150,000 at an annual rate of 6.0%. What are the (equal) monthly payments required to pay off the loan in 30 years?

8.6. Antibiotic doses. Every 12 hours Marcia takes a 200-milligram dose of antibiotic, which has a half-life of 24 hours. Find the amount of drug in Marcia's blood after the nth dose. Describe the steady state level of drug.

8.7. ◇ The pig problem. A pig weighing 200 pounds today gains 5 pounds per day with a food cost of 45 cents per day. The market price for pigs is 65 cents per pound today, but is falling 1 cent per day. How many days after today should the pig be sold to maximize the net profit?

8.8. Memory model. Suppose you are learning a foreign language. Each week you learn 25 new vocabulary words, but each week you also forget 5% of the words that you have already learned. How many words will you have learned after n weeks? What is the maximum number of words you can ever hope to learn?

8.9. ◇ Sleep model. Suppose that after many years of observation, you notice that your sleep is self-correcting: The amount of sleep you get one night is the average of the amount of sleep you got on the previous two nights. On two particular nights (initial conditions), you got 7 hours of sleep followed by 8 hours of sleep. How much sleep will you get n nights later? How much sleep will you get each night in the long run if this pattern continues?

8.10. Diluting a solution. Suppose you have a tank that is filled with 100 liters of a 40% (by volume) alcohol solution. You repeatedly perform the following operation: remove two liters of the solution in the tank and replace them with two liters of 10% alcohol solution. Let C_n be the concentration of the solution in the tank after the nth replacement.

(i) What is the concentration of the alcohol solution after the nth replacement?

(ii) After how many replacements does the alcohol concentration reach 15%?

(iii) What is the limiting concentration of the solution?

8.11. ◇ A Fibonacci property. Let F_n be the nth Fibonacci number. Show that

$$\lim_{n \to \infty} \frac{F_{n+1}}{F_n} = \frac{1 + \sqrt{5}}{2} \quad \text{(the golden mean)}.$$

8.12. The snowflake island. The snowflake island (or Koch island) is a fractal object that is the limit of repeatedly applying the following rule, beginning with an equilateral triangle.

> Trisect each side of the current figure, and build a new equilateral triangle on the middle third pointing outwards.

The figure below shows the second stage of the process. (i) What is the length of the coastline of the snowflake island? (ii) What is the area of the snowflake island?

8.13. ◇ China's one-son policy. In 1978, in an effort to reduce population growth, China instituted a policy that allows only one child per family. One unintended consequence has been that, because of a cultural bias toward sons, China now has many more young boys than girls. To solve this problem, some people have suggested replacing the one-child policy with a one-son policy: A couple can have children until they have a boy. Suppose that the one-son policy were implemented and that natural birth rates prevailed (half boys and half girls). Compare the number of children under the two policies.

8.14. Double glass. An insulated window consists of two parallel panes of glass with a small spacing between them. Suppose that each pane reflects a fraction p of the incoming light and transmits the remaining light. Considering all reflections of light between the panes, what fraction of the incoming light is ultimately transmitted by the window?

8.15. ◇ Bouncing ball. Suppose that a rubber ball, when dropped from a given height, returns to a fraction p of that height. How long does it take for a ball dropped from 10 meters to come to rest? You need the fact that, in the absence of air resistance, a ball dropped from a height h requires $\sqrt{2h/g}$ seconds to fall to the ground, where $g = 9.8$ meter/sec^2 is the acceleration of gravity. Also, the time taken to bounce *up* to a given height equals the time to fall from that height to the ground.

8.16. Random walk on the line. Imagine standing at the origin of a number line. On the toss of a fair coin, you take one step to the right (positive direction) if the coin shows a head, and you take one step to the left (negative direction) if the coin shows a tail. You repeat this process until you reach an exit at either the (integer) point $N < 0$ or the (integer) point $M > 0$, at which point the walk ends. Let p_k denote the probability that the walk will end at N if you are standing at the point k. If you move left at the next step, the probability that you reach the left exit is p_{k-1}, and if you move right at the next step, the probability that you reach the left exit is p_{k+1}; these moves have the same probability. Therefore the sequence of probabilities satisfies the difference equation

$$p_k = \frac{1}{2}(p_{k+1} + p_{k-1}) \quad \text{for} \quad k = N+1, \ldots, M-1.$$

We also know that $p_N = 1$ (if you are standing at the left exit, the probability is 1 that the walk ends at the left exit), and $p_M = 0$ (if you are standing at the right exit, the probability is 0 that the walk ends at the left exit).

(i) Solve the above difference equation, subject to the boundary conditions $p_N = 1$ and $p_M = 0$.

(ii) What is the probability that if you are standing at the origin, you will reach the left exit at N? Check the solution in the special case that $N = -M$. Is it consistent with your intuition?

8.17. ◇ **Habitat transitions.** A wildlife biologist knows from previous observations that each month 0.6 of the deer in the highland habitat remain in the highland habitat, while 0.4 of the deer in the highland habitat migrate to the lowland habitat. She also knows that each month 0.8 of the deer in the lowland habitat remain in the lowland habitat, while 0.2 of the deer in the lowland habitat migrate to the highland habitat. At the present time, 0.9 of the deer are in the highland habitat, and 0.1 of the deer are in the lowland habitat. Assume that these transition rates remain constant and that all deer of interest are currently in and remain in one of the two habitats.

(i) Find the solution that gives the fraction of the deer population in each habitat n months after the present, for $n = 0, 1, 2, \ldots$.

(ii) Find the steady state distribution of deer between the two habitats.

8.18. Stacking dominoes. Consider a set of identical dominoes that are one inch wide and two inches long. The dominoes are stacked on top of each other with their long edges aligned so that each domino overhangs the one beneath it. If there are n dominoes in the stack, what is the largest distance that the top domino can be made to overhang the bottom domino? How many dominoes can be stacked altogether before the stack topples?

8.3   Hints and Answers

8.1. HINT: Combine exponential growth and decay. ANSWER: The population is given by $P_n = 650(1.015)^n$, and it first exceeds 2500 when $n = 91$ years.

8.2. HINT: Find the supply and depletion difference equation. ANSWER: (i) The population is

$$P_n = 4000(1.015)^n - 80\frac{(1.015)^n - 1}{0.015}.$$

(ii) After 23 months, the population is $P_{23} = 3455$ fish. The population reaches zero at about 93 months. (iii) If the harvesting rate is no more than 60 fish per month, the population will increase.

8.3. HINT: Find the supply and depletion difference equation. ANSWER: (i) The balance is

$$B_n = 100\frac{(1.0075)^n - 1}{0.0075} \quad \text{for} \quad n = 0, 1, 2, \ldots.$$

(ii) It takes 123 months to reach a balance of $\$20,000$.

8.4. HINT: Find the supply and depletion difference equation. ANSWER: The loan balance is zero after 445 months.

8.5. HINT: Leave the monthly payment as a variable in the difference equation, solve the difference equation, and determine the loan payment. ANSWER: The required monthly payment is $p = \$899$.

8.6. HINT: With a half-life of 24 hours, a single dose is diminished by a factor of $2^{-t/24}$ in t hours. ANSWER: The amount of drug after the nth dose is

$$\frac{200\sqrt{2}(1 - a^n)}{\sqrt{2} - 1},$$

where $a = \frac{1}{\sqrt{2}}$. The steady state level of drug is $\frac{200\sqrt{2}}{\sqrt{2}-1}$.

8.7. HINT: Find the weight of the pig, the price per pound, the food cost, and the market value of the pig on the nth day. ANSWER: The pig should be sold after eight days.

8.8. HINT: Form a difference equation for the number of words learned and retained after n weeks.

ANSWER: The number of words learned and retained after n weeks is $500(1 - (0.95)^n)$. The maximum number of words you can hope to learn is 500 words.

8.9. HINT: Write a second-order difference equation for the amount of sleep on the nth night.

ANSWER: The amount of sleep on the nth night is $x_n = \frac{23}{3} - \frac{2}{3}\left(-\frac{1}{2}\right)^n$. In the long run, if the pattern persists, you will get close to $\frac{23}{3}$ hours of sleep per night.

8.10. HINT: Concentration is volume of alcohol per 100 liters of solution. Find the difference equation for the concentration.

ANSWER: (i) The concentration after the nth replacement is $C_n = 0.3 \cdot 0.98^n + 0.1$. (ii) The concentration first reaches 15% on the 89th day. (iii) The limiting concentration is 10%.

8.11. HINT: Divide the difference equation by F_n and assume that the limit exists.

8.12. HINT: Find general expressions for the length of each line segment, the number of line segments, the perimeter, and the area at stage n of the process, where $n = 0, 1, 2, \ldots$. Then let $n \to \infty$.

ANSWER: (i) The perimeter is infinite. (ii) The area is $\frac{2\sqrt{3}}{5}$.

8.13. HINT: Look at N families, and count the numbers of boys and girls born to these families.

ANSWER: With the one-son policy, there will be twice as many children as there are with the one-child policy.

8.14. HINT: How many times can a "ray" of light that is ultimately transmitted be reflected?

ANSWER: The fraction of light ultimately transmitted is $\frac{1-p}{1+p}$.

8.15. HINT: The sum of the rise and fall times is a geometric series.

ANSWER: The time to come to rest is $1.43\left(\frac{1+\sqrt{p}}{1-\sqrt{p}}\right)$ seconds.

8.16. HINT: Use a trial solution and apply the initial conditions.

ANSWER: (i) The probability is $p_k = \frac{M-k}{M-N}$. (ii) At the origin, $p_0 = \frac{M}{M-N}$. In the case that $M = -N$, $p_0 = \frac{1}{2}$.

8.17. HINT: Find the transition matrix and its eigenvalues.

ANSWER: (i) The fraction of deer in the highlands is $H_n = \frac{1}{3} + \frac{17}{30}0.4^n$. The fraction in the lowlands is $L_n = \frac{2}{3} - \frac{17}{30}0.4^n$. (ii) There will be twice as many deer in the lowlands as in the highlands.

8.18. HINT: Build the stack downward; that is, put the nth domino beneath the previous $(n - 1)$ dominoes.

ANSWER: The greatest possible length of the overhang with $n + 1$ dominoes is $\sum_{k=1}^{n} \frac{1}{k}$. The stack never topples, and the overhang can be made arbitrarily large.

Chapter 9
Insight and Computing

The purpose of computing is insight, not numbers.

— Richard Hamming

9.1 Pólya's Method for Computing

The purpose of this chapter is to illustrate the role of computing in problem solving. The topic is vast, and we cannot hope to do much more than highlight some typical problems and strategies. For the purposes of this chapter, we use *computing* inclusively to refer to everything from graphing calculators to high-powered desktop computers and work stations. While the use of computers may seem like an abrupt transition from the problem solving of the previous chapters, there is a plausible link; it is Pólya's method. Without too much modification, it is possible to reformulate Pólya's method so that it applies to problem solving with computers. Here is the computational version of Pólya's method, which we will use implicitly from now onward.

Pólya's Method for Computing

Step 1: Understand the problem. The first step is to determine where you are going. Be sure you understand what you are given (this might be input to a program) and what is asked (this might be output from the program).

Step 2: Plan a strategy for solving the problem. The next step is to decide the goal of the computation and how to reach it. As with problem solving, this step is the most difficult, and it requires creativity, organization, and experience. It can be made easier using *flowcharts* and *pseudocodes* (to be discussed shortly).

Step 3: Execute your strategy, and revise it if necessary. This step amounts to implementing your strategy on a computer, which might mean writing a program, doing numerical calculations, doing symbolic calculations, or drawing graphs. Revision of the strategy may be necessary, which requires returning to Step 2.

Step 4: Check and interpret your result. This step is absolutely critical on two levels. First, it is easy to write a program that runs but that does not do what you want it to do. Second, it is easy to write a program that does what you want it to do but that does not produce correct results! This step means checking that your program does what you want it to do *and* produces correct results. Once you feel confident in your program, then you can experiment and interpret the results.

There is no reason anyone would want a computer in their home.
— Ken Olson, president, chairman, and founder of Digital Equipment Corp., 1977

Example 9.1 Hailstone Numbers. We will now apply these ideas to a fascinating (unsolved) problem known as the *hailstone problem*, the Ulam conjecture, or the Collatz conjecture. Consider the following procedure:

1. Choose an integer $N \geq 2$ (the seed), and let it be the first term of a sequence a_0.

2. For $k = 0, 1, 2, 3, \ldots$, define the next term of the sequence as follows:

   - If a_k is even, then $a_{k+1} = a_k/2$.
   - If a_k is odd, then $a_{k+1} = 3a_k + 1$.
   - If $a_k = 1$, then STOP.

For example, with $N = 3$, the resulting sequence is $\{3, 10, 5, 16, 8, 4, 2, 1\}$, and the process stops after seven steps. Here are some interesting questions:

- For any starting number N, is this process guaranteed to stop (after a finite number of steps)?

- For a given starting number N, how many steps does it take for the process to stop?

Solution: Surprisingly, neither question has been answered. The Ulam or Collatz conjecture is that the process will eventually stop for any $N \geq 2$. The first problem we address is this: Given an integer $N \geq 2$, determine how many steps are required for the process to stop. We will explore this problem using the computer and Pólya's method. Here are the four steps.

1. **Understand the problem**. A program should accept a value of N (input) and determine how many steps are required for the process to stop (output). What if the process never stops? We should also have a way to tell the computer to stop if it executes some maximum number of steps, say 1000 steps, and has not reached a natural termination. Also, before writing a program, it might be a good idea to choose a few values of N and carry out the process by hand.

2. **Devise a strategy**. A **flowchart** is an outline of a computer program. It shows schematically what the program does, where it has to make decisions, and where it repeats instructions in a "loop." Figure 9.1 shows a possible

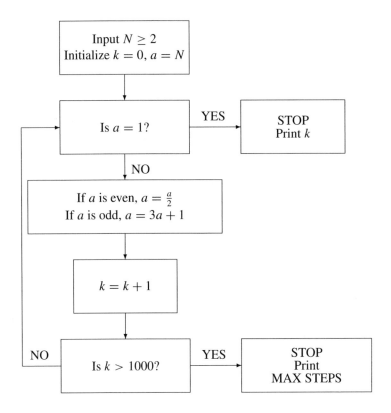

Figure 9.1. *A flowchart for the hailstone problem.*

flowchart for the problem at hand. Notice that k is a counter that is initialized with $k = 0$ and is incremented at each step of the process.

A **pseudocode** serves much the same organizational purpose as a flowchart; it is a step-by-step procedure written in a generic programming language. Here is a possible pseudocode for the hailstone problem.

Pseudocode for the hailstone problem

1. Input $N \geq 2$, the initial term of the sequence.
2. Initialize $k = 0$, $a = N$.
3. Enter the loop: While $a > 1$,
    Increment counter: $k = k + 1$.
    If a is even, $a = \frac{a}{2}$.
    If a is odd, $a = 3a + 1$.
    If $a = 1$, stop and print k.
    If $k > 1000$, stop and print MAX STEPS.

The statements $a = \frac{a}{2}$ and $a = 3a + 1$ that appear in both the flowchart and the pseudocode may seem peculiar. They should be interpreted as replacement

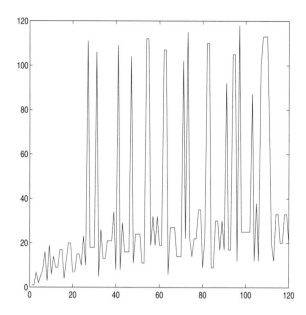

Figure 9.2. *The graph of hailstone numbers (number of steps to termination) vs. N shows that the process terminates for seeds N = 2, 3, ..., 120.*

statements: The first says *replace the current value of a by $\frac{a}{2}$*; similarly, the second statement says *replace the current value of a by* $3a + 1$. Some books would write these statements as $a \leftarrow \frac{a}{2}$ and $a \leftarrow 3a + 1$ to avoid multiple uses of the equal sign. Notice that by repeatedly overwriting a at every step, we don't see the individual terms of the sequence produced by the seed N. In testing this program, it is advisable to print the actual terms of the sequence to be sure they are correct.

3. **Carry out the strategy.** The procedure in the flowchart or pseudocode can now be implemented using the language or software package of your choice.

4. **Check, interpret, explain.** It is essential to check the program by comparing its output with hand calculations. Having developed some confidence in the program, you can begin experimentation with different values of N.

We can now build on the above program to test many values of N. Such a program would take a maximum value of N, call it N_{max}, and for each value of $N = 2, 3, 4, \ldots, N_{max}$, it would count the number of steps needed for the process to stop. The number of steps needed to stop (called a **hailstone number**) can be plotted as a function of N, as shown in Figure 9.2. The fact that all of the hailstone numbers are finite confirms the conjecture for $N \leq 120$. $\square$

Example 9.2 Approximating π. There are countless ways to approximate the irrational number π, many of which take the form of an infinite series. Here are two well-known series that define π:

$$\pi = 4 \sum_{k=0}^{\infty} \frac{(-1)^k}{2k+1} = 4\left(1 - \frac{1}{3} + \frac{1}{5} - \frac{1}{7} + \cdots\right),$$

$$\pi = 8 \sum_{k=0}^{\infty} \frac{1}{(4k+1)(4k+3)} = 8\left(\frac{1}{1 \cdot 3} + \frac{1}{5 \cdot 7} + \frac{1}{9 \cdot 11} + \cdots\right).$$

The goal is to use these two series to approximate π and to determine the accuracy of these approximations.

Solution: Once again, we follow Pólya's method.

1. **Understand the problem.** The two definitions of π involve infinite series. Because a computer can sum only a finite number of terms of a series, the best we can do is compute the partial sums of the series as approximations to π. So we will define the nth partial sums of the two series as follows:

$$\pi \doteq S_n = 4 \sum_{k=0}^{n-1} \frac{(-1)^k}{2k+1},$$

$$\pi \doteq T_n = 8 \sum_{k=0}^{n-1} \frac{1}{(4k+1)(4k+3)}.$$

2. **Devise a strategy.** The strategy will take the form of a flowchart or pseudocode for a program. The only user input we need is the number of terms, n, to be included in the partial sums. For output, we will display the nth partial sums, which are the approximations to π. If we have access to a high-precision approximation to π, then we might also want to display the error in the approximations. Here is a possible pseudocode for a program.

The sentence, *How I want a drink, alcoholic of course, after the heavy lectures involving quantum mechanics*, gives the first 15 digits of π. Do you see how?

Pseudocode for approximating π

1. Input n, the number of terms in the partial sums.
2. Initialize the partial sums: $s = 0$, $t = 0$.
3. Enter summing loop: For $k = 0$ to $n - 1$,
        Increment the first sum by the kth term:
        $$s = s + 4(-1)^k \frac{1}{2k+1}.$$
        Increment the second sum by the kth term:
        $$t = t + \frac{8}{(4k+1)(4k+3)}.$$
4. Exit loop and display the partial sums, s and t.
5. (Optional) Display the error in the sums, $|\bar{\pi} - s|$ and $|\bar{\pi} - t|$, where $\bar{\pi}$ is a high-precision approximation to π.

3. **Carry out the strategy.** The pseudocode can now be implemented in a program. With graphical output and a bit of elaboration, it is possible to display how the errors in the partial sums decrease with n (Figure 9.3). In order to produce these graphs, the program stores each partial sum $S_1, \ldots, S_n$ and $T_1, \ldots, T_n$, as well as the logarithm of the error in each partial sum. The

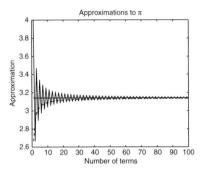
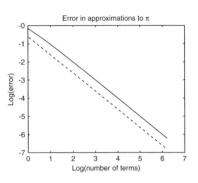

Figure 9.3. *The left plot shows approximations to π given by the first 100 partial sums of the two infinite series. The first series converges by oscillation (solid curve), and the second series converges monotonically from below (dashed curve). The right plot shows the logarithm of the error in the partial sums plotted against the logarithm of the number of terms.*

program ends by making graphs of the approximations to π and the logarithm of the error in the approximations.

4. **Check, interpret, explain**. The program should be checked with small values of n to be sure the first few partial sums are computed correctly. With increasing values of n, we expect to see better and better approximations to π. But how quickly do the approximations approach the exact value of π, and which series gives the better approximations?

The left plot of Figure 9.3 shows the first 100 partial sums of both series converging to the exact value of π indicated by the horizontal line. The partial sums of the first series converge by oscillation (solid curve), while the partial sums of the second series converge monotonically from below (dashed curve). The second series provides more accurate approximations to π for the same number of terms.

To get a sense of the speed of convergence, the right plot of Figure 9.3 shows the error in the first 500 partial sums of both series. It proves to be more informative to make a log-log plot of the error; that is, we plot log(error) against log n. With this perspective, we see that log(error) decreases linearly with log n. Furthermore, the slopes of both error lines are very nearly -1, which says that the error itself decreases roughly as n^{-1}. This turns out to be a very slow rate of convergence; the 500th partial sums provide only three correct digits of π. To obtain ten correct digits of π, roughly 5 million partial sums would be needed! Fortunately, more efficient methods for approximating π can be found (see Exercise 9.3). □

9.2 Root Finding

One of the most commonly occurring problems in mathematics is root finding: locating or approximating the solutions (roots) of an equation of the form $f(x) = 0$.

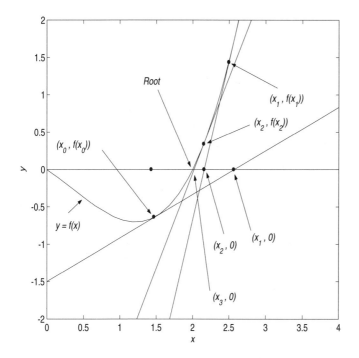

Figure 9.4. *The figure shows three steps of Newton's method applied to* $f(x) = \frac{1}{3}(x^3 - x^2 - 2x) = 0$ *with an initial approximation of* $x_0 = 1.5$. *The sequence* $\{x_0, x_1, x_2, \ldots\}$ *converges quickly to the root.*

With the symbolic and graphical power of today's software, this problem is not as formidable as it once was. For a wide variety of functions, many software packages can find roots symbolically. Often it suffices to use software to graph f and approximate roots.

Newton's method is the most widely used numerical procedure for root finding. In its simplest form, it is used to find the roots of a single differentiable function f. The method can be generalized to find roots of systems of nonlinear equations.

Newton's method is an **iterative** procedure, which means that we provide an initial approximation to a root and then successively improve the approximation. If all works as it should, the sequence of approximations generated by the method converges to a root. Newton's method is best derived geometrically, as shown in Figure 9.4. We have chosen an initial approximation of $x_0 = 1.5$. The first step is to find the point $(x_0, f(x_0))$ on the curve. Newton's method then approximates the curve $y = f(x)$ by the tangent line to the curve at $(x_0, f(x_0))$. The next approximation to the root, x_1, is the x-intercept of the tangent line. The process is then repeated.

We can now work out the form of the iteration explicitly. The equation of the tangent line (point-slope form of the line) at $(x_0, f(x_0))$ is

$$y - f(x_0) = f'(x_0)(x - x_0).$$

Isaac Newton, born in England in 1643, is best known for his contributions to physics and as a co-inventor of calculus. He was a professor at Cambridge University from 1669 to 1687. During the last 40 years of his life, he was a highly paid government official in London with little further interest in mathematical research. He died in London in 1727.

Setting $y = 0$, the x-intercept of the tangent line is

$$x_1 = x_0 - \frac{f(x_0)}{f'(x_0)}.$$

The process is repeated to generate the sequence of approximations $\{x_n\}$:

$$x_{n+1} = x_n - \frac{f(x_n)}{f'(x_n)} \quad \text{for} \quad n = 0, 1, 2, \ldots.$$

For the specific function $f(x) = x^2 - 1$, Newton's iteration takes the form

$$x_{n+1} = x_n - \frac{f(x_n)}{f'(x_n)} = x_n - \frac{x_n^2 - 1}{2x_n} = \frac{x_n^2 + 1}{2x_n} \quad \text{for} \quad n = 0, 1, 2, \ldots,$$

where the initial approximation x_0 is provided.

Figure 9.4 shows three steps of Newton's method applied to the equation $f(x) = \frac{1}{3}(x^3 - x^2 - 2x) = 0$, beginning with an initial approximation $x_0 = 1.5$. By the third iteration, the tangent lines approximate the curve very closely, and the approximations approach a root quickly. The first six approximations to the root are shown in Table 9.1. The sequence of approximations converges quickly to a root at $x = 2$.

Table 9.1. *The first six approximations to a root of $f(x) = \frac{1}{3}(x^3 - x^2 - 2x) = 0$ generated by Newton's method.*

n	x_n	n	x_n
0	1.5000000	3	2.0173396
1	2.5714286	4	2.002452
2	2.1580156	5	2.0000000

When Newton's method converges, it converges to a root and does so very quickly. However, convergence can be very sensitive to the choice of the initial approximation, x_0. If the initial approximation is too far from a root, the sequence generated by Newton's method may wander or diverge. Looking at the iteration formula, it's also clear that the method fails if $f'(x_n) = 0$ at any step.

The following pseudocode shows the basic features of a program for Newton's method. The program accepts an initial approximation, x_0, and a maximum number of iterations, I_{max}, as input. The function f and its derivative must also be defined by the user in such a way that they can be evaluated at any point. Finally, it is useful to terminate the iteration if the approximations are sufficiently close to a root. There are two common ways to test for convergence (although they do not always give an exact measure of the distance from a root): The *residual* for an approximation x_n is simply $f(x_n)$, which is small if x_n is close to a root. Similarly, if the difference between two approximations $|x_{n+1} - x_n|$ is decreasing, it is an indication that the sequence is converging. Here is a pseudocode for Newton's method that uses the residual for a convergence test.

Pseudocode for Newton's method

    1. Input x, the initial approximation, and I_{max}, the maximum number of iterations.

    2. Initialize counter $k = 0$ and termination cutoff, *tol*.

3. Define f and f'.
4. While $k \le I_{max}$ and $|f(x)| > tol$:
    a. Evaluate $p = f(x)$ and $q = f'(x)$,
    b. Compute the next approximation: $x = x - \frac{p}{q}$.
5. Print x, the last approximation to the root.

Example 9.3 The bowed rail revisited. Let's return to an earlier problem, to which we found an approximate solution. Recall the story: On a particularly hot day, a (hypothetical) railroad rail, originally $a = 1$ mile in length, expands by one foot (see Figure 9.5). Because the ends of the rail are anchored, the rail bows upward along the arc of a circle. How far off the ground is the midpoint of the bowed rail?

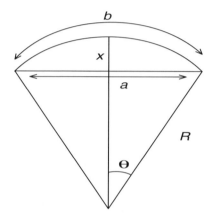

Figure 9.5. *A railroad rail of length $a = 1$ mile expands by one foot and bows upward along the arc of a circle. How far is the center of the rail from the ground? The figure is not drawn to scale.*

Solution: The goal is to derive the exact equations that describe the situation. Then several different strategies can be used to attack the equations. Note that $\frac{a}{2} = 2640$ feet is half the length of the original rail, and $\frac{b}{2} = 2640.5$ feet is half the length of the bowed rail. The rail is an arc of a circle of radius R, θ is half of the central angle, and x is the quantity we seek, the distance between the center of the rail and the ground. Three unknowns, R, θ, and x, have been introduced, so we need three equations. The relevant equations are

$$R\theta = \frac{b}{2} \quad (b \text{ is the length of arc of a circle}), \qquad (9.1)$$

$$R \sin\theta = \frac{a}{2} \quad (\text{right triangle relation}), \qquad (9.2)$$

$$(R - x)^2 + \left(\frac{a}{2}\right)^2 = R^2 \quad (\text{Pythagorean theorem}). \qquad (9.3)$$

The result is three nonlinear equations that cannot be solved exactly. However, some simplification can be done before resorting to approximations. Dividing the second equation by the first, we have

$$\frac{\sin\theta}{\theta} = \frac{a}{b} = 0.9998106419,$$

which leaves one equation in θ. If θ can be approximated from this equation, then the first equation determines R and the third equation determines x. So the immediate task is to solve

$$f(\theta) = \frac{\sin \theta}{\theta} - 0.9998106419 = 0. \tag{9.4}$$

Here are two approaches. Recall that $\lim_{\theta \to 0} \frac{\sin \theta}{\theta} = 1$, so the fact that $\frac{\sin \theta}{\theta}$ is so close to 1 tells us that θ is extremely small. For this reason, we can use a Taylor series to expand $\frac{\sin \theta}{\theta}$ about $\theta = 0$. Using the first two terms of the series, we can write

$$\frac{\sin \theta}{\theta} \doteq \frac{1}{\theta}\left(\theta - \frac{\theta^3}{6}\right) = 1 - \frac{\theta^2}{6}.$$

The approximation occurs because we terminated the series by neglecting a very small term proportional to θ^5. With this approximation, θ satisfies $1 - \frac{\theta^2}{6} = 0.9998106419$, which is easily solved to find that $\theta = 0.0337068013$.

Alternatively, we could apply Newton's method to (9.4). Table 9.2 shows the results, beginning with an initial approximation $\theta_0 = 0.5$.

Table 9.2. *The first ten approximations to a root of* $f(\theta) = \frac{\sin \theta}{\theta} - 0.9998106419 = 0$ *generated by Newton's method.*

n	θ_n	n	θ_n
0	0.50000000000000	5	0.03455036591817
1	0.24799856665706	6	0.03371803315289
2	0.12592151230569	7	0.03370776249964
3	0.06742929561752	8	0.03370776093528
4	0.04213555328953	9	0.03370776093527

After eight iterations, the sequence of approximations converges to $\theta = 0.033707$, which agrees with the Taylor series approximation. We now use (9.1) to find that $R = \frac{2b}{\theta} = 78{,}335.08$ feet and $x = R - \sqrt{R^2 - (a/2)^2} = 44.50$ feet. (The calculation of x involves subtracting two numbers of nearly equal magnitude, so it's advisable to carry as many significant digits as possible in the intermediate results.) We see that the rail is a surprising 44.50 feet above the ground. This result corroborates the value found in Example 3.7 using a straight-line approximation to the bowed rail. ☐

Example 9.4 Three ants in pursuit. We close with an example of a different use of computation, called **simulation**. Imagine three ants sitting at the vertices of an equilateral triangle with side lengths of one mile. At the same instant, each ant begins walking and continues walking at s miles per hour directly at the ant to its right. When do the ants collide, how far do they walk, and along what path do they walk?

Solution: It turns out that this problem can be solved analytically (see Chapter 11). However, there are many cases in which an analytical solution to a problem is either not obvious or impossible. When this occurs, a common strategy is to use a computer simulation to find approximate solutions. The three-ant problem is well suited to simulation. Here is how it works.

The goal is to write a program that tracks the movement of each ant. We have to follow all three ants because the motion of one ant determines the motion of the other two ants. The ants move continuously along smooth curves as they approach each other. However, a computer cannot store an infinite number of points, so it cannot exactly reproduce truly continuous motion. Instead, we imagine the ants moving in a sequence of many small steps. To do this, we use a *time step* Δt that can be chosen and varied by the user. At each time step, each ant finds the direction that it should walk (so that it walks directly toward its neighbor) and then takes a step of length $L = s\Delta t$ in that direction.

Only the curious will learn and only the resolute overcome the obstacles to learning. The quest quotient has always excited me more than the intelligence quotient.

— Eugene S. Wilson

A couple of observations are important. As the ants move, they continue to be at the vertices of an equilateral triangle that is rotating and shrinking. So the paths of the ants differ only by rotations of $\frac{2\pi}{3}$ radians (which is used in step 5a below). Also, at any moment of time the angle at which a particular ant moves is simply given by the slope of the line connecting that ant and its neighbor to the right (as expressed in step 5b below). Finally, note that as the ants approach one another, the distance between any pair of ants cannot be less than the step length L. So the test to *continue* the simulation is that the distance between any pair of ants exceeds L or that the step counter is less than a specified maximum. With those remarks, here is a pseudocode for the simulation.

Pseudocode for the three-ant problem

1. Input Δt, the time step; s, the speed of the ants; and k_{max}, the maximum number of steps.
2. Give the initial coordinates. Ant 1: $x_1(0) = 0$, $y_1(0) = 0$; Ant 2: $x_2(0) = 1$, $y_2(0) = 0$; Ant 3: $x_3(0) = \frac{1}{2}$, $y_3(0) = \frac{\sqrt{3}}{2}$.
3. Define $L = s\Delta t$, the length of a single step.
4. Initialize counter $k = 0$.
5. Begin simulation loop: While $k \leq k_{max}$ and the distance between two ants is greater than L,
   a. Find the directions, θ_i, in which the ants walk:

$$\tan \theta_1 = \frac{y_2(k) - y_1(k)}{x_2(k) - x_1(k)}, \quad \theta_2 = \theta_1 + \frac{2\pi}{3}, \quad \theta_3 = \theta_1 + \frac{4\pi}{3}.$$

   b. Advance each ant a distance L in the direction θ_i:

$$x_1(k+1) = x_1(k) + L \cos \theta_1, \quad y_1(k+1) = y_1(k) + L \sin \theta_1,$$
$$x_2(k+1) = x_2(k) + L \cos \theta_2, \quad y_2(k+1) = y_2(k) + L \sin \theta_2,$$
$$x_3(k+1) = x_3(k) + L \cos \theta_3, \quad y_3(k+1) = y_3(k) + L \sin \theta_3.$$

   c. Increment the step counter: $k = k + 1$.
6. Plot the points $(x_i(k), y_i(k))$ for $i = 1, 2, 3$.

Not surprisingly, the accuracy of the simulation depends critically on the time step Δt. For example, with $s = 1$ mile per hour, a time step of $\Delta t > 1$ would move each ant completely beyond the original triangle, and they would never meet. Also, as noted above, in order for the ants to approach a collision, the length of each step L must be small, which again means Δt must be small. Figure 9.6 shows the results of two simulations, one with $\Delta t = 0.01$ hours and one with $\Delta t = 0.005$ hours. The ant paths appear very similar in the two simulations, which is a good check on the time step. The first simulation required 69 steps, which means that ants traveled

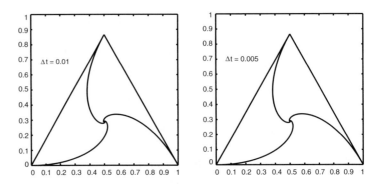

Figure 9.6. *The figure shows the simulation of the three-ant problem with time steps of* $\Delta t = 0.01$ *(left) and* $\Delta t = 0.005$ *(right). Ant 1 begins at* $(0, 0)$, *Ant 2 begins at* $(1, 0)$, *and Ant 3 begins at* $\left(\frac{1}{2}, \frac{\sqrt{3}}{2}\right)$.

for $69\Delta t = 0.69$ hours before meeting. The second simulation required 135 steps, so the ants traveled for $135\Delta t = 0.675$ hours before meeting; again, we have good agreement in the travel time. At a speed of $s = 1$ mile per hour, we can conclude that the ants traveled about 0.68 miles. Although these results are approximations, the simulations provided a lot of understanding about the problem.   □

The following exercises provide opportunities to write, test, and run short computer programs. In each case, follow Pólya's method as closely as possible. Specifically, write a flowchart or pseudocode for your programs and write an efficient, well-documented program to carry out the task. Try to judge the appropriate level of computation needed. Use analytical techniques when they are most efficient, but call on computation when necessary.

9.3 Exercises

9.1. ◇ **Comparing means**. Given n positive real numbers, $a_1, \ldots, a_n$, there are many ways to define their *average*. Here are three ways:

- arithmetic mean:

$$A = \frac{a_1 + \cdots + a_n}{n},$$

- geometric mean:

$$G = (a_1 \cdots a_n)^{1/n},$$

- harmonic mean:

$$\frac{1}{H} = \frac{1}{n}\left(\frac{1}{a_1} + \cdots + \frac{1}{a_n}\right).$$

Write and test a program to compute the three means of n arbitrary positive real numbers. Then carry out experiments to make a conjecture about the ordering of A, G, and H; that is, determine which one of the following inequalities is true: $A \leq G \leq H$, $A \leq H \leq G$, $H \leq G \leq A$, $H \leq A \leq G$, $G \leq H \leq A$, or $G \leq A \leq H$.

9.2. Sums of powers. Write a program that will accept a positive integer n for input and test the truth of the following identities:

$$\sum_{k=1}^{n} k = \frac{n(n+1)}{2},$$

$$\sum_{k=1}^{n} k^2 = \frac{n(n+1)(2n+1)}{6},$$

$$\sum_{k=1}^{n} k^3 = \frac{n^2(n+1)^2}{4} = \left(\sum_{k=1}^{n} k\right)^2,$$

$$\sum_{k=1}^{n} k^4 = \frac{n(n+1)(2n+1)(3n^2+3n-1)}{30}.$$

9.3. Approximations to π. Following Example 9.2, use the following three series to compute approximations to π. By experimenting with various partial sums, compare and discuss the rates of convergence of the three series. Which series provides the most efficient method for approximating π? For the first series, assume that cube roots can be evaluated exactly.

$$\frac{\pi^3}{32} = \sum_{n=0}^{\infty} \frac{(-1)^n}{(2n+1)^3},$$

$$\pi = \sum_{n=0}^{\infty} \frac{(-1)^n}{4^n}\left(\frac{2}{4n+1}+\frac{2}{4n+2}+\frac{1}{4n+3}\right),$$

$$\pi = 48\tan^{-1}\left(\frac{1}{18}\right) + 32\tan^{-1}\left(\frac{1}{57}\right)$$
$$-20\tan^{-1}\left(\frac{1}{239}\right),$$

$$\text{where} \quad \tan^{-1} x = \sum_{n=0}^{\infty} \frac{(-1)^n x^{2n+1}}{2n+1}.$$

9.4. Another form of 1. Consider the following representation of the number 1:

$$1 = 2\sum_{k=0}^{\infty} \frac{1}{(2k+1)(2k+3)}.$$

Write a program to approximate 1 using partial sums of this series. Based on experiments, determine how many terms are required to approximate 1 to at least five digits.

9.5. Generating primes. A prime number is an integer $p \geq 2$ whose only divisors are 1 and itself. Write a program that, given a positive integer n, lists all primes less than or equal to n. Here are two approaches:

(i) For each number k with $2 \leq k \leq n$, check to see if k has any divisors among $2, 3, 4, \ldots, K$, where K is the greatest integer less than or equal to $\sqrt{k}$; if not, call it a prime. Why can K be used as the largest test divisor?

(ii) Start with the list $\{2, 3, 4, \ldots, n\}$. Delete all multiples of 2 from the list; then delete all multiples of 3 from the list; continue in this manner using the smallest number that still appears on the list to delete all of its multiples from the list. The numbers that remain are the primes less than or equal to n. This method is attributed to the Greek Eratosthenes and is called the **sieve of Eratosthenes**.

Which of the two methods just proposed appears to be most efficient?

9.6. Goldbach conjecture. Christian Goldbach, born in Königsberg, Prussia (now Kaliningrad, Russia) in 1690, is best known for his conjecture, made in 1742 in a letter to Euler, that every even integer greater than 2 can be represented as the sum of two primes; the result has never been proved. The **Goldbach comet** illustrates how nearly the Goldbach conjecture is true. Given a positive integer N, let $4 \leq n \leq 2N$ be an even integer, and let $g(n)$ be the number of different ways that n can be expressed as the sum of two prime numbers. The graph of $(n, g(n))$ is a spray of points known as the Goldbach comet. If $g(n) = 0$ for any even integer, the conjecture fails. Write a program that generates the Goldbach comet for even integers between 4 and $2N$, where N is specified. [see "Goldbach's comet: The numbers related to Goldbach's conjecture," H. Fliegel and D. Robertson, *Journal of Recreational Mathematics*, 21(1), 1989]

9.7. ◇ Periodic savings plan simulation. Suppose that on your 25th birthday you begin making monthly deposits of $150 into an account that earns interest at a monthly rate of 0.5%. Assume that the balance in the account is initially zero.

(i) Write a program that simulates the investment process described. The program should take an age A as input and generate the balance in the account for every month until you reach age A. What is the balance when you reach age 45? age 65?

(ii) How do the balances at ages 45 and 65 change if the monthly interest rate is 0.7%? 0.3%?

(iii) If you could deposit 10% more each month, what is the percent increase in your balance at ages 45 and 65?

9.8. Loan simulation. Suppose that you take out a $150,000 loan for a new house. The interest rate for the loan is 6% per year, or 0.5% per month. Assume that in each month, interest based on the current balance is added to the balance, and you then make a payment of m that is subtracted from the balance.

(i) Write a program that simulates the loan process described. The program should take m as input and generate the loan balance for every month until the loan balance reaches zero. How long does it take to retire the loan if $m = $850? $m = $1050? $m = $1250?

(ii) What is the minimum monthly payment required to retire the loan (over any number of years)?

(iii) Estimate the monthly payment needed to retire the loan in 30 years.

9.9. ◇ The locker problem simulated. Five hundred students arrived for the first day of school and found 500 numbered lockers, all of which were closed. The first student switched every locker (opened closed lockers and closed open lockers); the second student switched every other locker (lockers $2, 4, 6, \ldots$). The nth student switched every nth locker (lockers $n, 2n, 3n, \ldots$). The pattern continued up to the 500th student. Write a program to simulate the process and describe the final state of the lockers. [*Pi Mu Epsilon Journal*, 1, April 1953, p. 350]

9.10. x^y versus y^x. Consider positive real numbers x and y for which $y \geq x$. Notice that $4^3 < 3^4$, while $3^2 > 2^3$ and $4^2 = 2^4$. Describe the regions in the first quadrant for which $y^x < x^y$ and $y^x > x^y$.

9.11. ◇ Tower of powers. For $a > 1$, how do you interpret

$$a^{a^{a^{a^{\cdots}}}} \quad ?$$

As it stands, the expression is ambiguous. The tower of powers can be built from the bottom or from the top. In other words, it could be evaluated as

$$x_{n+1} = a^{x_n} \quad \text{or} \quad x_{n+1} = x_n{}^a,$$

where $x_0 = a$. The two iterations have very different behaviors and depend critically on the value of a. Use all available techniques (analytical, graphing, numerical) to determine which definition gives finite values of $a^{a^{a^{a^{\cdots}}}}$. Using that definition, for what values of a is $a^{a^{a^{a^{\cdots}}}}$ defined, and what is its value?

9.12. Slicing a pizza. A 12-inch (diameter) pizza is cut into three pieces by making two parallel cuts, d inches to the right and left of the center of the pizza. What is the value of d such that the three pieces have the same area?

9.13. ◇ Pursuit problem. Imagine that a man is one mile east of his dog. At the same time, the dog starts walking north at 1 mile per hour and the man starts walking at 1.2 miles per hour towards his dog. At all times, the man walks directly toward his dog. Write a program to simulate this pursuit problem by alternately advancing the position of the man and the dog in time steps of Δt hours. Experiment with the choice of Δt, and use your calculations to estimate how much time elapses before the man catches his dog and how far the man walks.

9.14. Four ants in pursuit. Four ants sit at the vertices of a square with side lengths of one mile. At the same instant,

each ant begins walking at 2 miles per hour directly at the ant to its right. Write a program to simulate this pursuit problem by successively advancing the positions of the ants in time steps of Δt hours. Experiment with the choice of Δt, and use your calculations to approximate when the ants collide and how far they walk before colliding. (See Exercise 3.18.)

9.15. ◇ Ladders in the alley. Two ladders of length 20 and 25 feet criss-cross in an alley, extending from the foot of one wall to the opposite wall (see figure). If the intersection point of the ladders is six feet above the ground, how wide is the alley? Assume that the floor of the alley is horizontal and perpendicular to both walls.

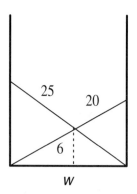

9.16. Exponential bike ride. Tom and Sue took a bike ride in which Sue gave Tom a head start of one hour. Tom started riding at 10 miles per hour, but fatigue decreased his speed by a factor of 4 every hour. Sue started riding at v_0 miles per hour, and her speed decreased by a factor of 2 every 2 hours.

(i) If Sue's initial speed was $v_0 = 5$ miles per hour, when and where did Sue catch Tom?

(ii) What is the minimum initial speed, v_0, at which Sue can start and still *eventually* catch Tom?

9.17. ◇ Ice cream cones. A sphere of ice cream is placed inside a right circular cone with a height of 1 unit and a base angle of $\frac{\pi}{6}$ radians. What is the radius of the ice cream sphere that maximizes the volume of ice cream below the top of the cone? [16]

9.18. Race strategy. Hank and Gloria are planning to run a race of length L. Gloria knows that George starts fast, at 8 miles per hour, but his running speed decreases with distance as $v(x) = 8e^{-x}$, where x is the distance from the start. Gloria would like to win the race, but also wants to run at a constant speed. If Gloria's maximum constant speed is 6 miles per hour, find the length of all races that Gloria can win with this strategy.

9.19. ◇ Another snowplow problem. With snow on the ground and falling at a constant rate, a snowplow began plowing down a long straight road at 12:00. The plow traveled *three* times as far in the first hour (between 12:00 and 1:00) as it did in the second hour (between 1:00 and 2:00). At what time did the snow start falling? Assume that the rate of plowing is inversely proportional to the cube of the snow depth.

9.20. Best speed function I. Suppose you plan to make a trip of length L miles and you begin driving at an initial speed of v_0 miles per minute. You have a choice of gaining speed in two ways: (A) you gain speed at a rate of 0.1 miles per minute every *minute* or (B) you gain speed at a rate of 0.1 miles per minute every *mile*. Find and compare the travel times with choices (A) and (B) (t_A and t_B) under the following conditions:

(i) You begin driving at $v_0 = 1.5$ miles per minute and the trip is $L = 50$ miles long.

(ii) You begin driving at $v_0 = 0.5$ miles per minute and the trip is $L = 15$ miles long.

(iii) You begin driving at $v_0 = 0.5$ miles per minute and the trip is $L = 50$ miles long.

(iv) What is the minimum initial speed v_0 such that (A) gives the faster travel time for all trip lengths?

(v) Challenge: Find the regions in the (v_0, L) parameter plane for which $t_A > t_B$ and $t_A < t_B$.

9.21. ◇ Best speed function II. Suppose you plan to make a trip of length L miles and you begin driving at an initial speed of v_0 miles per minute. You have a choice of losing speed in two ways: (A) you lose speed at a rate of one percent every *minute* or (B) you lose speed at a rate of one percent every *mile*. Find and compare the travel times with choices (A) and (B) (t_A and t_B) under the following conditions:

(i) You begin driving at $v_0 = 0.5$ miles per minute and the trip is $L = 40$ miles long.

(ii) You begin driving at $v_0 = 2$ miles per minute and the trip is $L = 100$ miles long.

(iii) You begin driving at $v_0 = 2$ miles per minute and the trip is $L = 190$ miles long.

(iv) What is the minimum initial speed v_0 such that (A) gives the faster travel time for all trip lengths?

(v) Challenge: Find the regions in the (v_0, L) parameter plane for which $t_A > t_B$ and $t_A < t_B$.

9.22. Running with a crosswind. Suppose that two runners start at the same point on a circular track and run a one-lap race. They both run at an average speed of 2 radians per minute, where one lap is 2π radians. The first runner

runs at a constant speed of $v_1 = 2$ radians per minute. The second runner's speed varies as $v_2(\theta) = 2 + \sin\theta$, where $0 \le \theta \le 2\pi$ (radians) is the position on the track (a good model for a crosswind). Who wins the race, and where do the runners pass each other?

9.23. Archimedes' π. The following procedure is based on an idea of Archimedes (although he certainly did not use a computer!). Consider a circle of diameter 1. If we can approximate its circumference (which is π), then we can generate approximations to π. We will use inscribed and circumscribed polygons to approximate the circumference of the circle.

(i) Suppose that we start with squares that are inscribed in and circumscribed around a circle of diameter 1. Let s_4 and t_4 be the side lengths of the circumscribed and inscribed squares, respectively. Let S_4 and T_4 be the perimeter of the circumscribed and inscribed squares, respectively. Find S_4 and T_4.

(ii) As shown in the figure below, suppose we now double the number of sides in the circumscribed and inscribed squares, to produce circumscribed and inscribed octagons. Let s_8 and t_8 be the side lengths of the circumscribed and inscribed octagons, respectively. Show that their perimeters, S_8 and T_8, are related to S_4 and T_4 by the expressions

$$S_8 = \frac{2S_4 T_4}{S_4 + T_4} \quad \text{and} \quad T_8 = \sqrt{T_4 S_8}.$$

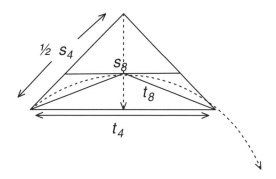

(iii) Explain why the same relations give S_{2n} and T_{2n} if we know S_n and T_n. That is, if we know the perimeter of the circumscribed and inscribed n-gons, then

$$S_{2n} = \frac{2S_n T_n}{S_n + T_n} \quad \text{and} \quad T_{2n} = \sqrt{T_n S_{2n}},$$

for $n = 4, 8, 16, 32, \dots$.

(iv) Write a program that starts with initial conditions S_4 and T_4 and generates S_{2n} and T_{2n} for $n = 4, 8, 16, 32, \dots$. Comment on how the accuracy of the approximations improves with each iteration. How many digits of accuracy do you gain with each

iteration? What is the best approximation to π that you can find using this method?

9.4 Hints and Answers

9.1. ANSWER: The ordering is $A \geq G \geq H$, with equality only if the n numbers are equal.

9.2. ANSWER: All of the identities are true.

9.3. HINT: Follow Example 9.2 and plot log(error) against $\log n$.

9.4. HINT: Follow Example 9.2 and plot log(error) against $\log n$.

9.5. HINT: Start with a flowchart or pseudocode.

9.6. HINT: Start with a flowchart or pseudocode.

9.7. ANSWER:

(i) With an interest rate of 0.5%, the balances are $69,306 and $298,724.

(ii) With an interest rate of 0.7%, the balances are $92,877 and $588,304. With an interest rate of 0.3%, the balances are $52,611 and $160,580.

(iii) All balances are 10% greater if the monthly deposit is 10% greater.

9.8. ANSWER:

(i) The loan will be retired after 430 months, 252 months, and 184 months, respectively.

(ii) The minimum monthly payment is $750.

(iii) A 30-year loan requires monthly payments of $899.33.

9.9. ANSWER: The open lockers are perfect squares $(1, 4, 9, 16, \ldots)$.

9.10. HINT: Find all points $y \geq x$ for which $y^x = x^y$. Then graph the regions in which $y^x < x^y$ and $y^x > x^y$.
ANSWER: The points for which $y^x = x^y$ lie on the curve given by $x = t^{1/(t-1)}$, $y = t^{t/(t-1)}$, where $t > 1$ is a parameter.

9.11. HINT: Determine whether and when the iterations $x_{n+1} = a^{x_n}$ and $x_{n+1} = x_n{}^a$ converge.
ANSWER: Building the tower from the top $(x_{n+1} = x_n{}^a)$ never leads to a finite value for $a > 1$. Building the tower from the bottom $(x_{n+1} = a^{x_n})$ converges to a finite value, provided $1 < a < e^{1/e} \doteq 1.445$. The value of the tower must be approximated numerically.

9.12. HINT: Use symmetry. The area of a sector of a circle with radius r is $A = \frac{1}{2}r^2\theta$. The length of a circular arc subtended by an angle θ is $s = r\theta$.

ANSWER: The pieces have equal size when $d = 1.59$ inches.

9.13. ANSWER: The man walks for approximately 2.72 hours or 3.26 miles.

9.14. ANSWER: Each ant walks one mile, which takes 0.5 hours.

9.15. HINT: Use similar triangles and the Pythagorean theorem.
ANSWER: The alley is 17.8 feet wide.

9.16. HINT: Find the velocities for both cyclists, then integrate to get the distance traveled.
ANSWER: (i) At $v_0 = 5$ miles per hour, Sue catches Tom after 2.95 hours or about 7.1 miles. (ii) Sue's initial speed must be more than 2.5 miles per hour in order to catch Tom.

9.17. HINT: A two-dimensional formulation suffices. Consider two cases: the center of the sphere above the top of the cone and the center of the sphere below the top of the cone.
ANSWER: A sphere with a radius of 0.216 units maximizes the amount of ice cream inside the cone.

9.18. HINT: Find the time needed for both runners to run a given distance.
ANSWER: If the race exceeds approximately 0.57 miles, then Gloria can win.

9.19. HINT: Let $t = 0$ be the time at which the snow started falling. Find the speed of the plow as a function of time, and integrate to find the distance traveled.
ANSWER: The snow started falling approximately 1.88 hours before 12:00.

9.20. HINT: Find the speed function for both cases and integrate to find the travel times. Then use graphs and numerical experiments to compare the travel times.
ANSWER: (i) (A) requires 14.7 minutes and (B) requires 20.0 minutes. (ii) (A) requires 13.9 minutes and (B) requires 13.0 minutes. (iii) (A) requires 23.0 minutes and (B) requires 27.0 minutes.

9.21. HINT: Find the speed function for both cases and integrate to find the travel times. Then use graphs and numerical experiments to compare the travel times.
ANSWER: (i) (A) requires 98.5 minutes and (B) requires 162.2 minutes. (ii) (A) requires 86.2 minutes and (B)

requires 69.5 minutes. (iii) (A) requires 254.0 minutes and (B) requires 233.7 minutes.

9.22. HINT: Write $\frac{d\theta}{dt} = 2 + \sin\theta$ or $\frac{dt}{d\theta} = (2 + \sin\theta)^{-1}$. The latter expression can be integrated with respect to θ.

ANSWER: The first runner (constant speed) wins and passes the second runner at approximately $\theta = 4.58$ radians (0.73 of a lap) and $t = 2.29$ minutes.

9.23. HINT: Use similar triangles to derive the relationships. Then note that $S_n = ns_n$ and $T_n = nt_n$.

Chapter 10

Take a Chance

> *Chance is always powerful. Let your hook be always cast; in the pool where you least expect it, there will be a fish.*
>
> — Ovid (43 B.C.E.–18 A.D.)

Entire books have been written about problem solving and probability [15, 27]. So in this single chapter devoted to probability, our goals must be somewhat modest. The first aim is to provide a *brief* introduction to the basic concepts of **discrete probability**, which is the name given to problems that involve a finite number of outcomes. As you will see, these problems can generally be solved using (often clever) counting methods. A far more detailed treatment of discrete probability can be found in any elementary probability and/or statistics book [6, 32, 38].

With the ideas of discrete probability in mind, we make a shift to a fascinating topic called **geometric probability**, which we approach on two fronts: analytical methods, which generally rely on calculus, and *random simulations*, which rely on the computer. This chapter weaves together many of the ideas and techniques introduced throughout this book.

10.1 Discrete Probability

Probability deals with situations in which chance or randomness determines which of several outcomes may occur. Tossing a coin results in one of two outcomes: heads or tails. Rolling a standard die results in one of six outcomes: 1, 2, 3, 4, 5, or 6. Drawing a card from a standard deck results in one of 52 outcomes. In each of these examples, we can assign to each outcome a probability that describes the likelihood that the outcome will occur. A probability is a number $0 \leq p \leq 1$, with the convention that $p = 0$ means an outcome is impossible and $p = 1$ means an outcome is certain.

In the case of tossing a fair coin, the probability of each outcome is 0.5, which we write as P(Head) $= 0.5$ and P(Tail) $= 0.5$. One interpretation of these probabilities is that if we toss a fair coin repeatedly, the proportion of heads (or tails) observed approaches 0.5 as the number of tosses increases.

In the case of rolling a fair die, the probability of any of the six outcomes is $\frac{1}{6}$. Specifically, we write

$$P(1) = P(2) = P(3) = P(4) = P(5) = P(6) = \frac{1}{6}.$$

And in the case of drawing a card randomly from a well-shuffled deck, the probability of drawing any specified card is $\frac{1}{52}$. For example, $P(3\heartsuit) = P(\text{King}\spadesuit) = \frac{1}{52}$.

When you have eliminated the impossible, whatever remains, however improbable, must be the truth.
— Sir Arthur Conan Doyle

Let's introduce some terminology with a more complex example. Suppose that we toss three fair coins. There are $2^3 = 8$ possible outcomes: HHH, HHT, HTH, HTT, THH, THT, TTH, TTT. The set of all possible outcomes is called the **sample space**. Clearly, each of the eight outcomes is equally likely, so we have, for example, $P(\text{HHT}) = \frac{1}{8}$. In many cases, we are interested in the probability of a certain set of outcomes, or **event**. For example, we might ask for the probability of tossing three coins and observing the event of two heads. Looking at the eight outcomes in the sample space, we see that two heads occur in three different outcomes: HHT, HTH, THH. Therefore $P(2 \text{ heads}) = \frac{3}{8}$.

To summarize, an event is a subset of the sample space consisting of one or more outcomes. When all outcomes are equally likely, the probability of event A is

$$P(A) = \frac{\text{number of outcomes in event A}}{\text{number of outcomes in the sample space}}.$$

This rule for computing the probability of an event will be used and generalized throughout this chapter. In more abstract terms, it says that the probability of an event is the size of the event relative to the size of the entire sample space.

Example 10.1 Coin probabilities. Find the probability of the following events when tossing four fair coins: 0 heads, 1 head, 2 heads, 3 heads, 4 heads.

Solution: We have $2^4 = 16$ outcomes in the sample space: HHHH, HHHT, HHTH, HHTT, HTHH, HTHT, HTTH, HTTT, THHH, THHT, THTH, THTT, TTHH, TTHT, TTTH, TTTT. Counting all instances of the various number of heads, we find the probabilities shown in Table 10.1. The table provides a **probability distribution** for all possible events. Notice that the probabilities in the table sum to 1, meaning that it is certain that either 0, 1, 2, 3, or 4 heads will appear.   □

Table 10.1. *Probability distribution for tossing four fair coins.*

Event	Probability
0 heads	$\frac{1}{16}$
1 head	$\frac{4}{16} = \frac{1}{4}$
2 heads	$\frac{6}{16} = \frac{3}{8}$
3 heads	$\frac{4}{16} = \frac{1}{4}$
4 heads	$\frac{1}{16}$
Total	1

Example 10.2 Card probabilities. In drawing a card at random from a well-shuffled deck of cards, what is the probability of drawing (i) a jack, (ii) an even-numbered card, (iii) a red royalty (jack, queen, king)?

Solution: In this problem, the sample space consists of the 52 cards in the deck. We must count the number of (equally likely) outcomes in each event. (i) Because there are four jacks in the sample space, we see that P(jack) $= \frac{4}{52} = \frac{1}{13}$. (ii) There are five even-numbered cards (2, 4, 6, 8, 10) in each of the four suits ($\spadesuit, \heartsuit, \diamondsuit, \clubsuit$), so there are 20 outcomes which give the event of drawing an even-numbered card. Thus, P(even-numbered card) $= \frac{20}{52} = \frac{5}{13}$. (iii) There are six cards that are red ($\heartsuit, \diamondsuit$) and royalty (jack, queen, king), so P(red royalty) $= \frac{6}{52} = \frac{3}{26}$.   □

So far we have computed probabilities simply by counting outcomes. In principle, this approach always works, but a few rules simplify matters considerably.

Important rules for computing probabilities

1. Two events A and B are **independent** if the probability of one event's occurring is not affected by whether the other event occurs. If A and B are independent, the probability that event A *and* event B occur is

$$P(A \text{ and } B) = P(A) \cdot P(B).$$

This result extends to multiple independent events. For example, in rolling a die three times, the outcome of one roll is independent of the other rolls. Therefore, the probability of rolling three 6's is

$$P(\text{three 6's}) = P(6 \text{ on 1st roll}) \cdot P(6 \text{ on 2nd roll}) \cdot P(6 \text{ on 3rd roll})$$
$$= \frac{1}{6} \cdot \frac{1}{6} \cdot \frac{1}{6} = \frac{1}{216}.$$

2. Two events are **mutually exclusive** or **disjoint** if they have no outcomes in common. If A and B are mutually exclusive, the probability that either event A *or* event B occurs is

$$P(A \text{ or } B) = P(A) + P(B).$$

This result extends to multiple mutually exclusive events. For example, in rolling two dice, the probability of rolling a sum of 5 is $\frac{1}{9}$, and the probability of rolling a sum of 7 is $\frac{1}{6}$ (verify these facts). These events are mutually exclusive, and so

$$P(\text{sum of 5 or 7}) = P(\text{sum of 5}) + P(\text{sum of 7}) = \frac{1}{9} + \frac{1}{6} = \frac{5}{18}.$$

For non–mutually exclusive events, we need to be more careful. In this case, we have

$$P(A \text{ or } B) = P(A) + P(B) - P(A \text{ and } B).$$

You can take as understood that your luck changes only if it is good.
— Ogden Nash

Chance favors only the prepared mind.
— Louis Pasteur

The extra term P(A and B) corrects for double counting the outcomes that belong to both A and B. For example, in drawing a card from a well-shuffled deck,

$$P(\text{queen or } \diamond) = \underbrace{\frac{1}{13}}_{P(\text{queen})} + \underbrace{\frac{1}{4}}_{P(\diamond)} - \underbrace{\frac{1}{52}}_{P(\text{queen and } \diamond)} = \frac{4}{13}.$$

3. The **complement** of an event A, denoted A', consists of all outcomes in the sample space that are not an instance of event A. In this case,

$$P(A') = 1 - P(A).$$

Notice that this rule implies that

$$P(A) + P(A') = 1,$$

which means it is certain that either A or its complement occurs. For example, the probability of rolling two dice and observing a sum *not* equal to 7 is

$$P(\text{sum not equal to 7}) = 1 - P(\text{sum equal to 7}) = 1 - \frac{1}{6} = \frac{5}{6}.$$

4. We can now investigate nonindependent events. The probability that event B occurs, *given that event A has occurred*, is called the **conditional probability** of B given A; it is denoted $P(B|A)$. A slight modification of the rule for independent events covers this case; we now have

$$P(A \text{ and } B) = P(A) \cdot P(B|A).$$

Notice that if A and B are independent, then $P(B|A) = P(B)$, and we recover the rule for independent events. For example, if we randomly select a jury pool from a group of 10 men and 10 women, the probability that the first two people selected are women is

$$P(2 \text{ women}) = P(\text{first a woman}) \cdot P(\text{second a woman} \mid \text{first a woman})$$
$$= \frac{10}{20} \cdot \frac{9}{19} = \frac{9}{38}.$$

The second term in this product $\left(\frac{9}{19}\right)$ reflects the fact that after one woman has been selected from the pool, 9 women and 19 people remain in the pool.

The conditional probability rule can be expressed in another useful form. It says

$$P(B \mid A) = \frac{P(A \text{ and } B)}{P(A)}.$$

For example, in rolling a single fair die, what is the probability of rolling a number less than 5, given that the roll is an even number? Letting event A be rolling an even number and event B be rolling a number less that 5, we have

$$P(\text{number} < 5 \mid \text{even number}) = \frac{P(\text{even number and less than 5})}{P(\text{even number})}$$
$$= \frac{\frac{1}{3}}{\frac{1}{2}} = \frac{2}{3}.$$

This result can also be verified by simple counting.

5. A common question asks for the probability that an event occurs *at least once* in a certain number of trials. For example, what is the probability that in rolling a fair die 10 times you observe at least one 6? Assuming that the events are independent, the simplest solution proceeds indirectly by first computing the probability of the complement event, that *no* 6 is observed in 10 rolls of the die. The probability that *no* 6 is observed in 10 rolls of the die is $\left(\frac{5}{6}\right)^{10}$. It follows that the probability of *at least* one 6 in 10 rolls of the die is $1 - \left(\frac{5}{6}\right)^{10} = 0.84$. To summarize, if the events are independent, the probability that event A occurs at least once in n trials is

$$P(A \text{ at least once in } n \text{ trials}) = 1 - (P(A'))^n.$$

In the problems considered so far, the required counting has been straightforward. A few additional counting techniques allow us to solve more difficult problems. Let's begin with the ideas known as **permutations** and **combinations**. Recall that, given n objects, there are $_nP_r$ permutations of those objects of size $r \leq n$, where

$$_nP_r = \frac{n!}{(n-r)!}.$$

The distinguishing property of permutations is that order matters; in other words, we count ABC, ACB, BAC, BCA, CAB, CBA as six different arrangements of A, B, and C.

On the other hand, given n objects, there are $_nC_r$ combinations of those objects of size $r \leq n$, where

$$_nC_r = \frac{n!}{r!(n-r)!}.$$

The distinguishing property of combinations is that order does not matter; in other words, we count ABC, ACB, BAC, BCA, CAB, CBA as one arrangement. The combinations formula uses the **binomial coefficients**, that is, $_nC_r = \binom{n}{r}$, pronounced *n choose r*.

Having introduced binomial coefficients, we should take one more step and consider problems that involve **Bernoulli trials**. Anytime we consider a process in which there are just two outcomes (heads/tails for a coin flip, boy/girl for births, hit/miss for shooting), it involves Bernoulli trials and binomial coefficients. For example, let's find the probability that an archer with a 20% bulls-eye average gets exactly three bulls-eyes in five shots. (A 20% average can be interpreted to mean that the probability of a bulls-eye on any shot is $p = 0.2$.)

There are many ways in which the archer can get three hits and two misses; for example, she may shoot HHHMM or MHMHH or HMHMH. Because the shots are independent, the probability of three hits and two misses in *any one of these orderings* is $p^3(1-p)^2$, where $p = 0.2$. So we need to count all of the ways in which three hits and two misses may occur. Think of the five shots as five slots, A, B, C, D, E, into which the three hits can be placed; for example, ABD means that the first, second, and fourth shots are hits (as does BAD), while CDE means that the third, fourth, and fifth shots are hits (as does EDC). Clearly the ordering of the slots makes no difference. So the number of ways that three hits can occur in five shots is the number of combinations of five objects taken three at a time, which is $\binom{5}{3}$. Each of these $\binom{5}{3}$ ways to have three hits has a probability of $p^3(1-p)^2$. Therefore the probability that the archer gets exactly three hits in five shots is

$$\binom{5}{3}p^3(1-p)^2 = 10 \cdot 0.2^3 \cdot 0.8^2 = 0.0512.$$

Generalizing this idea, we arrive at the rule for probabilities in Bernoulli trials: If the probability of success in one trial is p, the probability of having exactly k successes in n trials is

$$\binom{n}{k}p^k(1-p)^{n-k}.$$

Now let's put all of these ideas to work.

Example 10.3 Drawing balls from an urn. An urn contains six red balls, four green balls, and ten blue balls. If three balls are drawn at random without replacement, find the probability of the following events:

  (i) all three balls are blue,
 (ii) all three balls are green,
(iii) two are red and one is blue,
(iv) one of each color is drawn,
 (v) they are drawn in the order blue, green, red,
(vi) at least one is red.

Solution: It is important to note that balls are not put back into the urn after they are drawn, which means that the outcome of the second and third draws depends on the outcome of the previous draws.

  (i) One approach is to use conditional probabilities. Note that

$$P(3 \text{ blue balls}) = P(\text{blue on 1st draw})$$
$$\times P(\text{blue on 2nd draw} \mid \text{blue on 1st draw})$$
$$\times P(\text{blue on 3rd draw} \mid \text{blue on 1st and 2nd draws})$$
$$= \frac{10}{20} \cdot \frac{9}{19} \cdot \frac{8}{18} = \frac{2}{19}.$$

Alternatively, we can note that the number of ways that three blue balls can be selected from the set of ten blue balls is $\binom{10}{3}$. Similarly, the number of ways that three balls can be selected from the entire set of 20 balls (sample space) is $\binom{20}{3}$. Therefore, the probability of drawing three blue balls is

$$P(3 \text{ blue balls}) = \frac{\binom{10}{3}}{\binom{20}{3}} = \frac{2}{19}.$$

 (ii) Using the second technique in part (i), the probability of selecting three green balls is

$$P(3 \text{ green balls}) = \frac{\binom{4}{3}}{\binom{20}{3}} = \frac{1}{285}.$$

(iii) The number of ways to select two red balls (from six red balls) and one blue ball (from ten blue balls) is $\binom{6}{2} \cdot \binom{10}{1}$. Therefore, the probability of drawing two red balls and one blue ball is

$$P(\text{2 red balls, 1 blue ball}) = \frac{\binom{6}{2} \cdot \binom{10}{1}}{\binom{20}{3}} = \frac{5}{38}.$$

(iv) There are six ways to choose one red ball, four ways to choose one green ball, and ten ways to choose one blue ball. Therefore, the probability of drawing one ball of each color is

$$P(\text{one ball of each color}) = \frac{6 \cdot 4 \cdot 10}{\binom{20}{3}} = \frac{4}{19}.$$

(v) We can use conditional probabilities here:

$$\begin{aligned} P(\text{red, green, blue}) = {} & P(\text{first ball red}) \\ & \times P(\text{second ball green} \mid \text{first ball red}) \\ & \times P(\text{third ball blue} \mid \text{first ball red and second ball green}) \\ = {} & \frac{6}{20} \cdot \frac{4}{19} \cdot \frac{10}{18} = \frac{2}{57}. \end{aligned}$$

(vi) Using the rule for *at least one* events, we first note that there are 14 nonred balls. Therefore,

$$P(\text{none is red}) = \frac{\binom{14}{3}}{\binom{20}{3}} = \frac{91}{285}.$$

The probability of the complement event is

$$P(\text{at least one red}) = 1 - \frac{91}{285} = \frac{194}{285}. \qquad \square$$

Example 10.4 The famous birthday problem. What is the probability that with n people in a room, at least two have the same birthday?

Solution: We assume that all birthdays are equally likely and ignore leap year birthdays. We first compute the probability that n people have n different birthdays. The first of the n people has a birthday on a certain day with probability $\frac{365}{365} = 1$. The second person has a different birthday than the first person with a probability of $\frac{364}{365}$ (the birthday can fall on any of the other 364 days of the year). The probability that the third person has a birthday different than the first two people is $\frac{363}{365}$. Continuing in this manner, the probability that the nth person has a different birthday than the previous $n - 1$ people is $\frac{365-n+1}{365}$. Therefore, the probability that all n birthdays are different is

Even after correcting for the different lengths of months, births do not occur with the same frequency in each month. The U.S. birth rate is highest in September and lowest in January. On the subject of births, boys and girls are not born in equal proportions: 105 boys are born for every 100 girls.

$$\begin{aligned} P(n \text{ different birthdays}) &= \frac{365}{365} \cdot \frac{364}{365} \cdot \frac{363}{365} \cdots \frac{365 - n + 1}{365} \\ &= \left(1 - \frac{1}{365}\right)\left(1 - \frac{2}{365}\right) \cdots \left(1 - \frac{n-1}{365}\right). \end{aligned}$$

We seek the complement event that at least two people share a birthday. Therefore,

$$P(\text{shared birthday}) = 1 - \left(1 - \frac{1}{365}\right)\left(1 - \frac{2}{365}\right)\cdots\left(1 - \frac{n-1}{365}\right).$$

Table 10.2 shows the probability that among n people at least two people share a birthday, for $2 \le n \le 25$. The surprise in these results is that with only 23 people in the room, there is more than a "50-50" chance that at least two people have the same birthday.   □

Table 10.2. *Probability that at least two among n people have the same birthday.*

n	P	n	P
3	0.0082	15	0.2529
2	0.0027	14	0.2231
3	0.0082	15	0.2529
4	0.0164	16	0.2836
5	0.0271	17	0.3150
6	0.0405	18	0.3469
7	0.0562	19	0.3791
8	0.0743	20	0.4114
9	0.0946	21	0.4437
10	0.1169	22	0.4757
11	0.1411	23	0.5073
12	0.1670	24	0.5383
13	0.1944	25	0.5687

Example 10.5 Duelling Idiots. This problem gets its name from a superb probability book [27] that opens with the sentence, "This is a book for people who really like probability problems." Here is the problem, given in the same context, but somewhat more briefly. Two fellows, A and B, decide to duel with a six-shooter revolver. They put a single bullet in one of the six chambers and give the cylinder a spin. Then A shoots at B. If the gun fires, the game is over and A wins. If the gun does not fire, B takes the gun, gives the cylinder a spin and shoots at A. The duel continues in this manner until one idiot shoots the other. Here is the question: What is the probability that A, who shoots first, wins? Apart from the gruesome story, the problem has some valuable probability lessons.

Solution: As in many probability problems, the key to getting started is to identify the sample space. In this case we must look at all ways in which A can win. Letting Y mean that the gun fires and N mean that the gun does not fire, A can win with any of the following sequences of firings: Y, NNY, NNNNY, or any even number of Ns followed by a Y. Because these are mutually exclusive events, we can find the probability that A wins by summing the probabilities of each of these sequences. Before doing so, note that the probability of the gun firing on any round is $\frac{1}{6}$, because the gun has six randomly selected chambers, only one of which has the bullet. It follows that the probability of the gun not firing is $\frac{5}{6}$. Because the trigger pulls are independent events, the probability of, say, the NNNNY sequence is $\left(\frac{5}{6}\right)^4\left(\frac{1}{6}\right)$. Summing the probabilities of all of A's winning sequences gives us the probability of A winning the duel:

$$P(A) = \frac{1}{6} + \left(\frac{5}{6}\right)^2\frac{1}{6} + \left(\frac{5}{6}\right)^4\frac{1}{6} + \cdots.$$

Clearly the duel could last indefinitely, so the sum is infinite. Note that

$$6P(A) = \sum_{k=0}^{\infty}\left(\frac{5}{6}\right)^{2k},$$

where the series on the right is a geometric series. Recall that if $|a| < 1$, then

$$\sum_{k=0}^{\infty} a^k = \frac{1}{1-a}.$$

For our particular sum, $a = \left(\frac{5}{6}\right)^2$. Therefore

$$6P(A) = \sum_{k=0}^{\infty}\left(\frac{5}{6}\right)^{2k} = \frac{1}{1-\left(\frac{5}{6}\right)^2} = \frac{36}{11}.$$

It follows that $P(A) = \frac{6}{11} = 0.545454\ldots$. Because either A or B must eventually win, the probability of B winning is $1 - \frac{6}{11} = \frac{5}{11}$. We see that A has the better chance of winning, because A shoots first.   □

10.2 Geometric Probability

As mentioned at the outset, our goal is to approach geometric probability in two ways: using analytical (or *a priori*) methods and random (or Monte Carlo) simulations. Analytical methods will follow naturally from our previous work. Therefore we begin by introducing random simulations.

Most computers and software packages have a **random number generator** that produces numbers that are uniformly distributed on the interval $(0, 1)$; that is, each real number on $(0, 1)$ representable by the computer is equally likely to be selected. Random numbers are often needed on intervals other than $(0, 1)$, such as $(-1, 1)$ or $(0, 2\pi)$. We let `rand` be the command that generates a single random number on the interval $(0, 1)$ and let `floor(x)` produce the greatest integer less than or equal to x. You should verify that

The generation of random numbers is too important to be left to chance.
— Robert R. Coveyou

- `a*rand` produces a random number on the interval $(0, a)$,

- `rand + b` produces a random number on the interval $(b, b + 1)$,

- `c + (d-c)*rand` produces a random number on the interval (c, d), where $c < d$,

- `floor(2*rand)` produces a random sequence of 0's and 1's (good for simulating a coin toss),

- `floor(6*rand + 1)` produces a random sequence of 1's, 2's, 3's, 4's, 5's, and 6's (good for simulating the roll of a die), and

- `floor(n*rand + m)` produces integers from the set $\{m, m+1, \ldots, m + n - 1\}$.

Example 10.6 Throwing dart to approximate π. Let's begin with an example that is easy to visualize. Imagine a circle of radius $\frac{1}{2}$ inscribed in a unit square (see Figure 10.1). Suppose you throw darts at the board and that each dart is guaranteed to land inside the square, but its landing position within the square is completely random (any position is as likely to occur as any other position). What is the probability that a dart will land inside the circle?

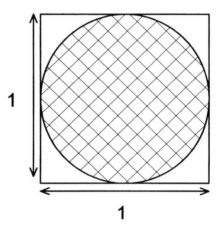

Figure 10.1. *A square dart board with a circular target can be used to approximate π.*

Solution: The analytical approach relies on the fundamental rule that we have used throughout the chapter to compute discrete probabilities:

$$P(A) = \frac{\text{number of outcomes in event A}}{\text{number of outcomes in the sample space}}.$$

In this case, event A is the dart landing inside the circle, while the sample space is all possible locations of the dart inside the square. The new feature in this problem is that we cannot count the number of outcomes in event A or in the sample space, because the dart can land in an (uncountably) infinite number of locations; this is no longer a problem in discrete probability. However, the basic rule generalizes in a natural way. The number of outcomes in event A corresponds to the area of the circle, and the number of outcomes in the sample space corresponds to the area of the square. Therefore with geometric probability problems, we will use the analogous rule

$$P(A) = \frac{\text{size of event A region}}{\text{size of the sample space}}.$$

In many problems, *size* means *area*; but in more general problems, *size* could mean *volume* or higher-dimensional measures. For our dart board problem, we have

$$P(\text{dart lands in circle}) = \frac{\text{area of circle}}{\text{area of square}} = \frac{\pi \left(\frac{1}{2}\right)^2}{1} = \frac{\pi}{4}.$$

We see that the theoretical probability that a randomly thrown dart lands within the circle is $\frac{\pi}{4} = 0.785$.

Using a simulation approach, we could approximate this probability by randomly throwing many darts at the board and counting the number of throws and the number of hits (darts that land inside the circle). The ratio of hits to throws should give an approximation to $\pi/4$.

Never tell people how to do things. Tell them what to do and they will surprise you with their ingenuity.
— George S. Patton, Jr.

The dart-throwing experiment is most easily done with a computer simulation. In this case, we call a random number generator to determine the (x, y) coordinates of the dart, where $0 < x < 1$ and $0 < y < 1$. It is then a simple matter to determine whether the dart lands inside of the circle. Because the center of the circle is $\left(\frac{1}{2}, \frac{1}{2}\right)$, the dart lands inside the circle if

$$\sqrt{\left(x - \frac{1}{2}\right)^2 + \left(y - \frac{1}{2}\right)^2} < \frac{1}{2} \quad \text{or} \quad \left(x - \frac{1}{2}\right)^2 + \left(y - \frac{1}{2}\right)^2 < \frac{1}{4}.$$

A pseudocode that carries out this procedure is given below. It takes the number of throws as input, produces random values of x and y in the interval $(0, 1)$, counts the hits, and computes an approximation to π.

Pseudocode for random darts

1. Input N, the number of throws.
2. Set $M = 0$ to count hits.
3. For $k = 1$ to N
    $x = \text{rand}, \ y = \text{rand}.$
    If $(x - 0.5)^2 + (y - 0.5)^2 < 0.25$, then $M = M + 1$.
4. Compute approximate value of π as $\frac{M}{N}$.

While every simulation produces a different approximation to π, we expect that, in general, the more throws used in a simulation, the better the resulting approximations. Table 10.3 shows the approximations to π generated in six different simulations with an increasing number of throws. Monte Carlo simulations are notoriously slow to converge, meaning that a large number of trials is needed to generate good approximations. On the other hand, simulations are very inexpensive, so in terms of overall computing cost, simulations can be competitive with other methods. □

Example 10.7 The random chord problem. Suppose you randomly choose two points on the unit circle (radius 1) and connect them with a chord. What is the probability, $p(r)$, that the length of the chord is less than a specified number $0 \le r \le$

Table 10.3. *Circular dart board approximations to π.*

Number of throws	Approximation to π	Error in approximation
1000	3.0720	0.0696
2000	3.1540	0.0124
4000	3.1500	0.0084
6000	3.1300	0.0116
8000	3.1555	0.0139
10,000	3.1352	0.0064

2? Clearly, $p(0) = 0$, $p(2) = 1$, and p increases monotonically. But what does the graph of p look like on the interval $0 \leq r \leq 2$?

Solution: The analytical approach allows us to compute the probability exactly, as follows. First note that the problem is symmetric in the sense that we do not need to select two points on the circle randomly. As shown in Figure 10.2 (left) we can fix one point arbitrarily, say at $(1, 0)$. Then we can choose the second point randomly and draw a chord of length L between the points. The location of the second point is given by the angle θ, which can be randomly generated. It suffices to consider values of θ in the interval $0 \leq \theta \leq \pi$.

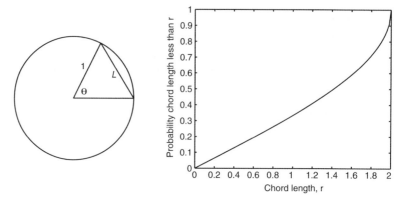

Figure 10.2. *A random chord of length L is created by a random angle θ (left). The probability, $p(r) = \frac{2}{\pi} \sin^{-1} \left(\frac{r}{2} \right)$, that the length of a random chord has length less than r is plotted as a function of r (right).*

A short calculation gives us the relationship between L and θ. The perpendicular bisector of L passes through the center of the circle and bisects θ. The length of the chord subtended by θ is

$$L(\theta) = 2 \sin \left(\frac{\theta}{2} \right).$$

We see that the random chord has length less than r, provided that

$$2 \sin \left(\frac{\theta}{2} \right) < r \quad \text{or} \quad \theta < 2 \sin^{-1} \left(\frac{r}{2} \right).$$

For example, the random chord has length less than $r = \sqrt{3}$, provided $\theta < \frac{2\pi}{3}$, and the random chord has length less than $r = 1$, provided $\theta < \frac{\pi}{3}$. To find the probability, $p(r)$, that the random chord has length less than r, we take the ratio of the values of θ for which the chord has length less than r to all possible values of θ, which in this case is π; that is,

$$p(r) = \frac{\text{values of } \theta \text{ for which the chord has length less than } r}{\text{all possible values of } \theta}$$

$$= \frac{2\sin^{-1}\left(\frac{r}{2}\right)}{\pi}$$

$$= \frac{2}{\pi}\sin^{-1}\left(\frac{r}{2}\right).$$

The resulting graph of $p(r)$ is shown in Figure 10.2 (right). Notice that $p(0) = 0$ and $p(2) = 1$, as expected.

A pseudocode for a simulation to approximate $p(r)$ is given below. The number of trials and the value of r are specified by the user. Within the simulation loop, one point on the circle is fixed at $(1, 0)$, and the second point is determined by the angle $0 \le \theta < 2\pi$. The distance between the two points, L, is then computed and compared to r. If $L < r$, then the hit counter is incremented. An approximation to $p(r)$ is found by computing the ratio of "hits" to the total number of trials.

Pseudocode for random chords

1. Input N, the number of throws, and r, the maximum chord length.
2. Fix one point at $x_1 = 1$ and $y_1 = 0$.
3. Set $M = 0$ to count hits.
4. For $k = 1$ to N
   $\theta = 2\pi *$rand,
   $x_2 = \cos\theta,\ y_2 = \sin\theta,$
   $L = \sqrt{(x_1 - x_2)^2 + (y_1 - y_2)^2}.$
   If $L < r$, then $M = M + 1$.
5. Compute the approximate value of $p(r)$ as $\frac{M}{N}$.   □

Example 10.8 Buffon needle problem. One of the classical problems in geometric probability is the Buffon needle problem. Imagine a flat surface scored with parallel lines one unit apart. A needle of unit length is dropped randomly onto the surface. What is the probability that the needle lands on a line?

Solution: The theoretical approach uses Figure 10.3 (left) to model the situation. The position of the needle can be specified by two variables: the distance between the midpoint of the needle and one of the vertical lines, x, and the angle of the needle, θ. By symmetry, it suffices to consider $0 \le x \le \frac{1}{2}$ and $0 \le \theta \le \frac{\pi}{2}$. Note that the horizontal extent of the needle is $\frac{1}{2}\cos\theta$. Thus, if $\frac{1}{2}\cos\theta > x$, then the needle crosses the vertical line, and we have a *hit*.

The space of all possible needle positions is the region $0 \le x \le \frac{1}{2}$ and $0 \le \theta \le \frac{\pi}{2}$ (see Figure 10.3 (right)). The curve that determines whether the needle hits or misses is $x = \frac{1}{2}\cos\theta$. Needle positions below this curve correspond to hits, and

At the age of 20, Georges Louis Leclerc Comte de Buffon (1707–1788) discovered the binomial theorem. During his lifetime, he made contributions to mechanics, geometry, probability, number theory, and differential and integral calculus. His first book introduced calculus into probability theory.

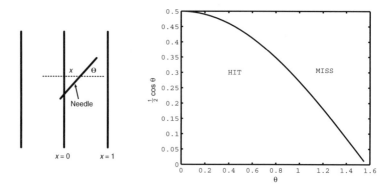

Figure 10.3. *The geometry of the needle problem; a hit occurs if $\frac{1}{2}\cos\theta > x$ (left). The space of all possible needle positions is divided into the HIT region and the MISS region by the curve $x = \frac{1}{2}\cos\theta$ (right).*

needle positions above this curve correspond to misses. Thus the probability of a hit is the ratio of the area under the curve to the area of the entire sample space of positions, or

$$p = \frac{\int_0^{\pi/2} \frac{1}{2}\cos\theta\, d\theta}{\frac{\pi}{2} \cdot \frac{1}{2}} = \frac{2}{\pi}.$$

The needle problem provides an unexpected experimental way to approximate π; however, it is inefficient!   □

In the following problems, attempt to find an analytical solution. Simulations should be used in the absence of an analytical solution or to corroborate an analytical solution.

10.3 Exercises

10.1. ◇ **Selecting a quartet**. Of the 16 students at a party, 4 are from Colorado (CO), 5 are from Utah (UT), and 7 are from Ohio (OH). Four students are selected at random from this group of 16 students. Find the following probabilities.

(i) All four students are from OH.

(ii) Two of the four students are from CO and two are from UT.

(iii) At least one of the four students is from OH.

(iv) All four students are from one state.

(v) At least one of the students is from each state.

10.2. Drawing balls. Ten balls numbered from 1 to 10 are placed in a large box. Three balls are drawn randomly without replacement.

(i) What is the probability that the three balls are numbered 5 or greater?

(ii) What is the probability that the three balls are all odd-numbered?

(iii) What is the probability that the three balls are drawn in the order odd, even, odd?

(iv) What is the probability that the three balls are drawn in sequential order (for example, 3, 4, 5)?

(v) What is the probability that the three balls are drawn in increasing order (for example, 3, 7, 9)?

10.3. ◇ **Poker probabilities**. Poker players are dealt five cards from a well-shuffled standard deck of cards. Find the probability of being dealt the following hands:

(i) a full house (three of a kind and a pair; for example, three kings and two tens),

(ii) four of a kind (for example, four kings),

(iii) three of a kind (for example, three queens) and no better,

(iv) a pair (for example, 2 aces) and no better,

(v) a straight (five cards in sequence of any suit; for example, 3, 4, 5, 6, 7) and no better, which excludes a straight flush (all of the same suit),

(vi) a flush (five cards of the same suit) and no better, which excludes a straight flush (all of the same suit).

10.4. Shooting free throws. A basketball player has a 70% free throw shooting average, which can be interpreted to mean that the probability of his hitting any single free throw is 0.7. Assume that free throws are independent events.

(i) What is the probability that he makes his first five shots?

(ii) What is the probability that in shooting five free throws he has at least one hit?

(iii) What is the probability that in shooting ten free throws he has exactly eight hits?

(iv) What is the probability that he gets his first hit on the third shot?

(v) What is the probability that in shooting ten free throws he has a streak of exactly five consecutive hits?

10.5. ◇ Die rolling. A fair die is rolled ten times. What is the probability of rolling at least a five at least twice?

10.6. Three strange coins. Of three coins, one has two heads, one has two tails, and one has one head and one tail. You draw one coin randomly from a bag, look at *only one side* of the coin, and see a head. What is the probability that the coin has two heads?

10.7. ◇ Royalty at the top. What is the probability that at least one of the top three cards in a well-shuffled deck of cards is a jack, queen, or king? [20]

10.8. Defects. The defect rate in an assembly line for computer chips is 3%; that is, on average 3 out of every 100 chips are defective. Out of 10,000 chips produced in a day, a random sample of 500 chips is chosen. What is the probability that exactly 10 of the 500 chips in the sample are defective?

10.9. ◇ Dice tossing game. Two people play a game in which they alternately toss a pair of dice. The first person to get a sum of 7 on the dice wins the game. Find the probability that the one who tosses first wins the game. [38]

10.10. The first probability problem. Probability is said to have originated when a seventeenth-century French gambler, Chevalier de Mére, played the following game: He bet his rivals even money that if he rolled two fair dice 24 times he would see at least one double-six. Over the long run, he lost money at this game, which defied his intuition. In search of an explanation, he wrote to Pascal, a leading mathematician of the day, who in collaboration with Fermat provided an explanation. What is the probability of rolling at least one double-six in 24 rolls of a pair of dice?

10.11. ◇ Mammogram probability. A recent health report claimed that if a woman has an annual mammogram for ten years in a row, the probability of having at least one false report during those ten years is 0.50. What is the probability of a false report on a single mammogram?

10.12. Underdogs in the World Series. Assume that the strongest team in the World Series has a probability $p > \frac{1}{2}$ of winning any particular game. What is the probability that the weaker team wins the World Series? (The World Series is won by the first team to win four games.) [27]

10.13. ◇ Odd person out. Suppose that N people go out for coffee, with the agreement that the odd person out on a joint coin flip pays the bill. In other words, each of the N people flips a coin, and if exactly one person gets a head (and $N - 1$ people get a tail) or if exactly one person gets a tail (and $N - 1$ people get a head), the odd person pays. If there is no odd person, the coin toss is repeated. With N people in the group, what is the average number of tosses needed to determine an odd person out? (The average number of tosses is a weighted average $\sum_k k p_k$, where p_k is the probability that an odd person is determined in k coin tosses.) [27]

10.14. Letter-envelope problems. Suppose you write 23 different letters and address 23 different envelopes for the recipients. If you randomly put one letter into each envelope, what is the probability that

(i) exactly one letter is in the wrong envelope,

(ii) exactly two letters are in the wrong envelope, and

(iii) exactly three letters are in the wrong envelope?

10.15. ◇ Sets of random numbers. Suppose that the command `rand` produces (uniformly distributed) random numbers on the interval $(0, 1)$. Explain how to use `rand` to produce a random number from the following sets:

(i) a random lottery number from the given set $\{1, 2, 3, 4, \ldots, 41, 42\}$,

(ii) a random real number on the interval $(3, 6)$.

10.16. Two random numbers. Two random numbers are selected from the interval $(0, 1)$. What is the probability that the sum of their squares in less than 1?

10.17. ◇ **Two random numbers**. Two random numbers are selected from the interval $(0, 1)$. What is the probability that the distance between them is less than $\frac{1}{2}$?

10.18. Roots of quadratics. Suppose that the real numbers b and c are taken randomly from the interval $(-a, a)$, and consider the equation $x^2 + bx + c = 0$.

   (i) What is the probability that the roots of the equation are real when $a = 1$?

   (ii) What is the probability that the roots of the equation are real for arbitrary values of a? Comment on the existence of real roots as $a \to \infty$.

10.19. ◇ **Random chords: Bertrand's paradox**. An equilateral triangle is inscribed inside of a unit circle. Imagine generating a random chord of the circle in two ways: (A) randomly choose two points on the unit circle and connect them or (B) randomly choose a point Q inside the unit circle and draw the shortest possible chord that has Q as a midpoint. Using (A) and (B), what is the probability that the length of the chord exceeds the side length of the equilateral triangle? The fact that (A) and (B) produce two different probabilities is called Bertrand's paradox. In fact, other definitions of *random chord* lead to different probabilities.

10.20. Generalized Buffon needle problem. A flat surface is scored with parallel lines d units apart. What is the probability that a needle, one unit in length, randomly dropped on the surface, will land on a line when (i) $d > 1$ and (ii) $d < 1$? Check that both results are consistent with the probability for $d = 1$ found in Example 10.8.

10.21. ◇ **Triangular dart board**. A circle is inscribed in an equilateral triangle. What is the probability that a dart, which is equally likely to land at any point inside the triangle, lands inside the circle?

10.22. Inscribed rectangle. Let C be the unit circle $x^2 + y^2 = 1$. A point P is randomly chosen on the circle, and another point Q is chosen randomly in the interior of C. Let R be the rectangle with sides parallel to the x- and y-axes with diagonal PQ. What is the probability that no point of R lies outside of C?

10.23. ◇ **Triangle sides**. Given a random triangle, what is the probability that the length of at least one side is less than or equal to the arithmetic mean of the other two side lengths? [20]

10.24. Edge vs. center. A randomly thrown dart hits a square target. What is the probability that the dart lands nearer the center of the square than to any edge of the square?

10.25. ◇ **Chord of a square**. Two points are randomly chosen on the edge of a square. What is the probability that the length of the line joining the two points is less than the length of a side of the square?

10.26. Another chord of a square. Two points are chosen randomly from adjacent sides of a square. What is the probability that the area of the triangle formed by the sides of the square and the line between the two points is less than one-fourth of the area of the square?

10.27. ◇ **Points in a semicircle**. Three points are chosen randomly inside a circle. What is the probability that there is a semicircle (of the original circle) in which all three points lie? Equivalently, what is the probability that the triangle formed by the three points contains the center of the circle?

10.28. Chance of meeting. A young man and woman plan to meet between 5:00 and 6:00 p.m., each agreeing not to wait more than 10 minutes for the other person. Find the probability that they will meet if they arrive independently at random times between 5:00 and 6:00.

10.29. ◇ **Breaking a stick**. A one-meter stick is broken at two random points along its length. What is the probability that the resulting three sticks can be arranged to form a triangle?

10.30. Random triangle. Two line segments AB and CD have lengths 8 and 6, respectively. Points P and Q are selected randomly on AB and CD, respectively. What is the probability that the area of a triangle with height AP and base CQ is greater than 12? [38]

10.31. Points on the unit interval. On the interval $[0, 1]$, n points are chosen randomly.

   (i) ◇ Find the probability that the point lying farthest to the right will be to the right of 0.6.

   (ii) Find the probability that the point lying farthest to the left will be to the left of 0.6.

   (iii) Find the probability that the point second from the left will be to the left of 0.6.

10.32. Approximating e. Suppose that the unit interval is subdivided into N bins of width $1/N$. Now generate N random numbers on the unit interval, and put each number into the appropriate bin. [27]

   (i) Consider this process as a set of Bernoulli trials, and compute the probability, $p_N(n)$, that a certain bin receives exactly $n \le N$ of the random numbers.

   (ii) What is the probability, $p_N(0)$, that a certain bin receives exactly $n = 0$ random numbers? Evaluate $p_\infty(0) = \lim_{N \to \infty} p_N(0)$.

   (iii) Write a simulation program that takes the number of bins N as input and approximates $p_N(0)$. By increasing N, approximate $p_\infty(0)$, and hence e.

10.33. Overlapping chips. Imagine a circle of unit radius drawn on a flat surface. Two poker chips of radius r, where $0 < r < \frac{1}{2}$, are thrown randomly so that their centers land within the circle. Write a simulation program to approximate the probability that the two poker chips land overlapping each other. You should imagine an ideal situation in which the poker chips do not bounce or roll; they land precisely where they are thrown. Is an analytical solution possible?

10.34. Random walks. Imagine standing at the origin of a number line. On the toss of a fair coin, you take one step to the right (positive direction) if the coin shows a head, and you take one step to the left (negative direction) if the coin shows a tail. You repeat this process until you reach an exit at either the (integer) point $N < 0$ or the (integer) point $M > 0$, at which point the walk ends. (See Exercise 8.16 for an analytical approach to this problem.)

(i) Write a program that takes $N < 0$ and $M > 0$ as input and simulates the random walk, assuming that you begin at the origin. The program should generate the sequence of positions $\{x_k\}$, where x_k is your position after the kth move; note that $x_0 = 0$. The program should terminate when either $x_k = N$ or $x_k = M$.

(ii) Perform 500 simulations to estimate the probability that you reach the left exit in the case that $N = -M$.

(iii) Perform 500 simulations to estimate the probability that you reach the left exit in the general case that $N \neq -M$.

(iv) Perform 500 simulations to estimate the average number of steps required to reach either exit in the case that $N = -M$.

10.4 Hints and Answers

10.1. HINT: Count carefully the number of ways each event can occur and the total number of outcomes.
ANSWER: (i) $\frac{1}{52}$, (ii) $\frac{3}{91}$, (iii) $\frac{121}{131}$, (iv) $\frac{41}{1820}$, (v) $\frac{1}{2}$.

10.2. HINT: Count carefully the number of ways each event can occur and the total number of outcomes.
ANSWER: (i) $\frac{1}{6}$, (ii) $\frac{1}{12}$, (iii) $\frac{5}{36}$, (iv) $\frac{1}{15}$, (v) $\frac{1}{6}$.

10.3. HINT: In all cases, there are $\binom{52}{5} = 2{,}598{,}960$ different five-card hands. Account for four different suits and thirteen different card values $(2, 3, \ldots, J, Q, K, A)$.
ANSWER: (i) 0.00144, (ii) 0.000240, (iii) 0.0211, (iv) 0.423, (v) 0.00392, (vi) 0.00197.

10.4. HINT: Use probability rules for independent events and for Bernoulli trials.
ANSWER: (i) 0.168, (ii) 0.998, (iii) 0.233, (iv) 0.0630, (v) 0.00204.

10.5. HINT: Consider the complement events.
ANSWER: 0.896.

10.6. HINT: Count the number of successes within the sample space.
ANSWER: $\frac{2}{3}$.

10.7. HINT: Compute the probability of the complement.
ANSWER: 0.553.

10.8. HINT: Use Bernoulli trials.
ANSWER: 0.0479.

10.9. HINT: The first person can win with any of the sequences W, LLW, LLLLW, ..., where W is a win and L is a loss.
ANSWER: $\frac{6}{11}$.

10.10. HINT: Use the "at least once" rule.
ANSWER: 0.491.

10.11. HINT: Use the "at least once" rule.
ANSWER: 0.0670.

10.12. HINT: A series can be won in 4, 5, 6, or 7 games. The winning team must win the last game.
ANSWER: $(1-p)^4(1 + 4p + 10p^2 + 20p^3)$.

10.13. HINT: Use Bernoulli trials and derive a sum of the form $\sum_{k=1}^{\infty} ka^{k-1}$, which has a value of $\frac{1}{(1-a)^2}$.
ANSWER: The expected (average) number of coin tosses is $2^{N-1}/N$.

10.14. HINT: Part (i) requires no calculation. Parts (ii) and (iii) require the number of ways that two and three letters can be mismatched with their envelopes.
ANSWER: (i) 0, (ii) 9.79×10^{-21}, (iii) 1.37×10^{-19}.

10.15. ANSWER: (i) `floor(42*rand + 1)`, (ii) `3 + 3*rand`.

10.16. HINT: Find the area of the success region in the unit square.
ANSWER: $\frac{\pi}{4}$.

10.17. HINT: Find the area of the success region in the unit square.
ANSWER: $\frac{3}{4}$.

10.18. HINT: Find the region of the (b, c) plane that produces real roots.
ANSWER: (i) $p(1) = \frac{13}{24}$; (ii) $p(a) = \frac{1}{2} + \frac{a}{24}$ for $0 < a \leq 4$ and $p(a) = 1 - \frac{2}{3\sqrt{a}}$ for $a \geq 4$; $\lim_{a \to \infty} p(a) = 1$.

10.19. HINT: Refer to Example 10.6 for random chord (A). For (B), find the area of the region within the unit circle that produces the required chord lengths.
ANSWER: (A) $\frac{1}{3}$, (B) $\frac{1}{4}$.

10.20. HINT: Follow Example 10.7, and use a modified sample space.
ANSWER: (i) $\frac{2}{d\pi}$, (ii) $\frac{4}{d\pi}\left(\frac{d}{2}\cos^{-1}d + \frac{1}{2}(1 - \sqrt{1-d^2})\right)$.

10.21. HINT: Find the ratio of areas.
ANSWER: $\frac{\pi}{3\sqrt{3}} \doteq 0.605$.

10.22. HINT: Find the region within which Q must lie in order to satisfy the conditions.
ANSWER: $\frac{4}{\pi^2} \doteq 0.405$.

10.23. HINT: No calculation is needed.
ANSWER: The probability is 1.

10.24. HINT: Find the curve along which points are equidistant from the center and the nearest edge.   Use symmetry.
ANSWER: $\frac{4\sqrt{2}-5}{3} \doteq 0.219$.

10.25. HINT: Draw a good picture, use symmetry, and assume that the first point lands on one fixed side of the square.
ANSWER: $\frac{1}{4} + \frac{\pi}{8} \doteq 0.643$.

10.26. HINT: Assume that the square is a unit square. Let the two random points have coordinates $(x, 0)$ and $(0, y)$.

ANSWER: $\frac{1}{2}(1 + \ln 2) \doteq 0.847$.

10.27. HINT: What is the probability that three random points inside a circle lie on the same side of a diameter of the circle?
ANSWER: $\frac{3}{4}$.

10.28. HINT: The arrival times (measured in fractions of an hour after 5:00) are random points within the unit square.
ANSWER: $\frac{1}{4}$.

10.29. HINT: What are the conditions on the side lengths of any triangle?
ANSWER: $\frac{1}{4}$.

10.30. HINT: Find the area of the region in the sample space that corresponds to triangles with an area greater than 12.
ANSWER: $\frac{1}{2}(1 - \ln 2) \doteq 0.153$.

10.31. HINT: Begin with two or three points and look for the pattern.

(i) ANSWER: $\frac{1}{n}(1 - 0.6^n)$,

(ii) ANSWER: $\frac{1}{n}(1 - 0.4^n)$.

(iii) ANSWER: $\frac{0.4^{n-1}}{n(n-1)}(0.6n + 0.4)$.

10.32. HINT: Recall that $e^x = \lim_{N\to\infty}\left(1 - \frac{x}{N}\right)^N$.

ANSWER: (i) $p_N(n) = \binom{N}{n}\left(\frac{1}{N}\right)^n\left(1 - \frac{1}{N}\right)^{N-n}$, (ii) $p_N(0) = \left(1 - \frac{1}{N}\right)^N$ and $p_\infty(0) = e^{-1}$.

Chapter 11

Toward Modeling

Mathematics is a way of thinking that can help make muddy relationships clear. It is a language that allows us to translate the complexity of the world into manageable patterns.

— K.C. Cole, *The Universe and the Teacup*

The word *modeling* means different things to different people. Imagine how the term would be interpreted by a mathematician, a statistician, an architect, and a fashion designer. By one interpretation of the word, this entire book has been about modeling: It is the process of translating a practical problem into a mathematical problem, and then solving it. This chapter, devoted entirely to exercises, is different from previous chapters only because the problems are somewhat more elaborate and are not identified with any particular method of solution. Many of the ideas and strategies of the previous chapters can be put to use. In no particular order, here are ten final problems.

11.1 Exercises

11.1. ◇ **Optimal bike riding**. Here is a simple model for riding a bicycle that has a selection of gears. Assume the bike has 21 gears, with gear ratios $\{1, 1.1, 1.2, \ldots, 3.0\}$. This means that if you ride in a gear with ratio r, then one full turn of the pedals advances the bike wheels by r revolutions; $r = 1$ is the lowest gear and $r = 3.0$ is the highest gear. Assume also that the bike has 27-inch wheels (diameter). As with all bikes, more energy is required to pedal at a given frequency with higher gears than with lower gears, so assume that each gear requires $p = 6\%$ more energy than the previous gear, where the lowest gear requires 1 unit of energy per revolution of the pedals.

(i) Suppose you wish to ride 1000 yards. Which gear should you use to minimize the total energy expended? How much energy do you use?

(ii) Suppose you have 1500 units of energy to spend. Which gear should you use to ride as far as possible? How far can you ride?

11.2. The tape deck problem. Most cassette or VCR players have a counter that shows the number of revolutions of the take-up reel as a tape is played. The counter does not advance at a uniform rate. Suppose that the radius of the hub of the take-up reel is r_0 inches, the tape moves over the play/record head at a constant rate of s inches per second, and the thickness of the tape is h inches. Find the relationship between the counter reading and the time elapsed.

11.3. ◇ **Draining a soda can**. A cylindrical soda can has a radius of 3 centimeters and a height of 12 centimeters. It is evident that when the can is filled with water, its center of gravity is at the midpoint of the long axis of the can. As the can is drained, the center of gravity moves downward along the long axis. However, when the can is completely drained, the center of gravity is once again at the midpoint of the long axis of the can. Find the lowest point of the center of gravity of the can as the can is drained. Neglect the mass of the can itself, and assume that the can is filled only with soda (density $\rho_\ell = 1$ gram per cubic centimeter) and air (density $\rho_a = 0.001$ gram per cubic centimeter).

11.4. Coding and information content. Imagine a coding system in which messages can be formed from a short signal (S) with a duration of one time unit and a long signal (L) with a duration of two time units. For example, a message of the form SSLSS, would have a duration of six time units. Let M_n be the number of different ways of forming a message of duration n. For example, $M_4 = 5$ since the five messages SSL, LL, SSSS, SLS, and LSS all have a duration of four time units.

(i) Find a general expression for any term of the sequence $\{M_n\}$.

(ii) The capacity of a channel is defined (by Claude Shannon, the innovator of information theory) to be

$$C = \lim_{n \to \infty} \frac{\log_2 M_n}{n}.$$

Find the capacity of the channel carrying messages of the form described above.

11.5. Lottery strategies. At last you won a staggering $10 million in the lottery. As with most lotteries, you have a choice of receiving your prize in one of two ways: With option A, often called the annuity option, you receive the full amount of the jackpot in n equal annual payments ($\$\frac{10}{n}$ million per year). With option B, often called the lump sum option, you receive a fraction f of the jackpot immediately. Which is best? Suppose that with either option, you plan to withdraw $100,000 per year for living expenses and invest the remaining available cash in an account that earns a constant $p = 5\%$ per year compounded annually. This means that with option A, you make annual deposits and, with option B, you deposit all of the funds immediately. Assume you win the lottery on January 1 and that the withdrawal of living expenses begins at the start of each year, immediately after interest is compounded. The first withdrawal occurs on the day you win. The first interest payment is at the end of the first year. [suggested by Corey Ayala, University of Colorado at Denver]

(i) ◇ If the lump sum option pays $f = 0.6$ of the total jackpot and the annuity option pays for $n = 10$ years, which option should you choose for the maximum gain after 10 years?

(ii) ◇ If the lump sum option pays $f = 0.6$ of the total jackpot and the annuity option pays for $n = 20$ years, which option should you choose for the maximum gain after 20 years?

(iii) For a fixed interest rate p, find and plot the break-even curve for the two options in the (n, f) plane; that is, if you are given a certain payout period of n years, what is the critical value of f that determines which option is best? Does this result depend on either the amount of the annual withdrawals or the amount of the jackpot?

(iv) Modify the above model to include a 5% per year cost-of-living increase in the annual withdrawals.

11.6. Cooling coffee. A fairly accurate model to describe the cooling of a conducting object is Newton's law of cooling. It says that if an object has an initial temperature of T_0 degrees and is placed in surroundings with an ambient temperature of A degrees, then the temperature of the object in time is given by $T(t) = A + (T_0 - A)e^{-kt}$, where $t \geq 0$ and the rate constant k is determined by the thermal properties of the object. The function does what you might expect: $T(0) = T_0$, $\lim_{t \to \infty} T(t) = A$, and the difference between the initial temperature and the ambient temperature decreases exponentially. Now suppose that q ounces of milk at temperature T_m are added to Q ounces of coffee at temperature T. A reasonable assumption is that the temperature of the coffee is immediately lowered to the average of the temperatures, weighted by the volumes; that is, the new temperature is

$$T_{new} = \frac{qT_m + QT}{q + Q}.$$

Suppose you have 8 ounces of 180-degree coffee in an 80-degree room and want to cool it to 115 degrees in 10 minutes. When should you add 2 ounces of 40-degree milk? Assume that $k = \frac{\log 2}{10}$.

11.7. ◇ **Master-dog pursuit problem**. At the moment a dog begins walking north from a crossroads at one mile per hour, a man begins walking from a point one mile east of the crossroads. The man walks at a constant speed of $s > 1$ miles per hour and at all times walks directly at the dog.

Where does the man meet the dog? (This problem requires the solution of separable differential equations.)

11.8. Caesar's last breath. This problem is generally attributed to the English physicist Sir James Jeans (1877–1946). The question is this: What is the probability that the next breath you take contains a molecule of air that was exhaled in the last breath of Julius Caesar (100–44 B.C.E.)? Clearly, we need some additional facts and assumptions. Because Caesar lived long ago, assume that the molecules of his last breath still exist and have been thoroughly mixed throughout the atmosphere, which itself consists of uniformly mixed molecules. A very rough estimate is that the atmosphere consists of 10^{44} molecules. A basic law of chemistry asserts that 22.4 liters of any gas contain 6×10^{23} molecules (Avogadro's number). Finally, assume for convenience that a breath of air contains 0.37 liters.

11.9. ◇ The reservoir problem. Imagine a large water reservoir shaped like an inverted frustrum of a pyramid (a square-based pyramid whose top has been sliced off along a plane parallel to the base). The horizontal cross-sections of the reservoir are squares that decrease in area from 225 square meters at the top of the reservoir to 100 square meters at the bottom. The depth of the reservoir is 10 meters. Let $h(t)$, $S(t)$, and $V(t)$ denote the depth, surface area, and volume of the water in the reservoir, respectively, at time $t \geq 0$. Assume that the reservoir loses water due to evaporation at a rate that is proportional to the area of the exposed water in the reservoir; that is, $V'(t) = -\alpha S(t)$, where $\alpha = 0.05$ meters/day. Find the volume of water in the reservoir at all times $t \geq 0$, assuming that the reservoir is full at $t = 0$. At what time will the reservoir be empty? (This problem requires the solution of a separable differential equation.)

11.10. Great snowplow chase. One day it started snowing at a constant rate. Three identical snowplows started out from the same place, at noon, 1:00 p.m., and 2:00 p.m. and began plowing down a long straight road. All three plows collided with each other at the same time. Assuming that plowing rates are inversely proportional to the snow depth, determine when the snow started falling. (This problem requires the solution of differential equations.) [attributed to Murray Klamkin, University of Alberta]

11.2   Hints and Answers

11.1. HINT: Find the number of pedal strokes needed to go a given distance in a given gear. Then find the energy required.
ANSWER: (i) Energy is minimized in the eighth gear, and the amount of energy expended is 375 units. (ii) Distance is maximized in the eighth gear, and the distance traveled is 4000 yards.

11.2. HINT: The radius of the tape on the take-up reel increases linearly with the number of revolutions. Let θ measure the angular movement of the take-up reel, n be the counter, and ℓ be the length of tape on the take-up reel. In an increment of time dt, $d\ell = s\, dt = r\, d\theta = 2\pi r\, dn$.
ANSWER: $t(n) = \frac{2\pi n}{s}\left(r_0 + \frac{1}{2}nh\right)$.

11.3. HINT: Assume that the can is weightless. Turn the can on its side, and find the balance point using the teeter-totter principle: If a person of mass m_1 is located at a coordinate x_1 and another person of mass m_2 is located at a coordinate x_2, the balance point (center of gravity) has the coordinate $\bar{x}$, where $m_1(x_1 - \bar{x}) = m_2(x_2 - \bar{x})$.
ANSWER: The low point of the center of gravity is 0.368 centimeters above the bottom of the can.

11.4. HINT: Show that the sequence $\{M_n\}$ satisfies the difference equation for the Fibonacci sequence.

ANSWER: The channel capacity is $\log_2\left(\frac{1+\sqrt{5}}{2}\right)$.

11.5. HINT: Write and solve difference equations for the two options.
ANSWER: (i) With the lump sum option, the balance after 10 years is $8.3 million. With the annuity option, the balance after 10 years is $12.8 million. (ii) With the lump sum option, the balance after 30 years is $18.8 million. With the annuity option, the balance after 30 years is $16.5 million.

11.6. HINT: Find the temperature of the coffee, $T(t)$, just prior to adding the milk at $t = t^*$. Then find the temperature of the coffee immediately after adding the milk. Then find the temperature of the coffee for $t > t^*$. Determine t^* so that $T(10) = 115$.
ANSWER: The milk should be added after about 3.2 minutes.

11.7. HINT: Draw a good picture, and use the fact that the tangent to the man's path points at the dog at all times.
ANSWER: The coordinates of the point of encounter are $\left(0, \frac{s}{s^2-1}\right)$.

11.8. HINT: Find the probability that a molecule of your next breath does *not* contain a molecule of Caesar's last breath. Use the fact that $\lim_{x\to\infty}\left(1 - \frac{1}{x}\right)^x = e^{-1}$.
ANSWER: The probability is approximately 0.63.

11.9. HINT: Draw a good figure of the cross-section of the reservoir at an arbitrary time $t > 0$.

ANSWER: The volume is given by $V(t) = (C - \frac{1}{3}at)^3 - V_L$, where $V_L \doteq 666.7$ and $C \doteq 13.10$. The reservoir is empty after 199.6 hours.

11.10. HINT: Determine the position of each plow successively.

ANSWER: The snow started falling at 11:30.

Chapter 12

Solutions

Chapter 1

1.9 Suppose that Reuben's birthday is December 31 and that he was speaking on January 1, 2004. Then two days ago (December 30, 2003) he was 20 years old. On December 31, 2003, he turned 21; on December 31, 2004, he would turn 22; and on December 31, 2005 (which is later next year), he would turn 23.

1.11 The boat and the ladder rise with the tide, so the number of rungs showing does not change.

1.13 The worst and best players are twins. However, the worst player's twin (who is the best player) and the best player are of the opposite sex, which is impossible.

1.15 There are four intervals between the five chimes, so each interval is $\frac{5}{4}$ second long (assuming that the chimes occur instantaneously). There are nine intervals between ten chimes, so it takes $9 \cdot \frac{5}{4} = \frac{45}{4} = 11.25$ seconds to strike 10:00.

1.17 The number of red cards in the first half of the deck is 26 minus the number of black cards in the first half of the deck. However, the number of black cards in the second half of the deck is also 26 minus the number of black cards in the first half of the deck. So the number of red cards in the first half of the deck equals the number of black cards in the second half of the deck, regardless of how the deck is shuffled.

1.19 If A is from Truth, he would tell the truth and say he is from Truth. If A is from Lies, he would lie and say he is from Truth. So A must have said he is from Truth. Therefore, B lied, and C told the truth. C is from Truth.

1.21 The facts that the brown book is separated from the pink book by two books and that the pink book is not last lead to three cases, which are shown in the table. The facts that the gold book cannot be first and that either gray-orange-pink or pink-orange-gray must appear eliminate all but the ordering brown, gray, orange, pink, gold.

pink			brown	
	pink			brown
brown			pink	

1.23 There is no missing dollar: The desk clerk has $25, the bellboy has $2, and each of the three guests has $1, which adds up to $30. It is a mistake to conclude that the guests spent $9 each for the room. With $2 in the bellboy's pocket, the cost of the room was effectively $28/3 = \$9\frac{1}{3}$ per person. After receiving $1 from the bellboy, each guest has effectively spent $\$8\frac{1}{3}$ for the room.

Chapter 2

2.1 Let r be the radius of the balls, which is also the radius of the can. Then the height of the can is $6r$ and the circumference of the can is $2\pi r$. Because $2\pi \doteq 6.28 > 6$, the circumference of the can is greater than the height of the can.

2.3 (i) In the worst case he could select 12 brown socks and then two black socks. He must select 14 socks to be sure of getting a black pair. (ii) In the worst case he could select a sock of either color before getting a pair of either color on the third draw. He must select three socks to be sure of getting a pair of either color.

2.5 It takes 9 digits to number the 9 pages between 1 and 9. It takes 180 digits to number the 90 pages between 10 and 99. It takes 2700 digits to number the pages between 100 and 999. It takes $4(n - 999)$ digits to number the pages between 1000 and n. Letting n be the number of pages in the book, we have

$$9 + 180 + 2700 + 4(n - 999) = 2989.$$

Solving for n, we see that the book has $n = 1024$ pages.

2.7 Let a be the man's age at death. The problem tells us that

$$\frac{a}{4} + \frac{a}{5} + \frac{a}{3} + 13 = a.$$

Solving for a, we see that $a = 60$.

2.9 Let p be the old price per dozen in cents. Relating the number of apples bought before and after the price change, we have

$$\frac{50}{p/12} = \frac{50}{(p - 10)/12} - 5.$$

Simplifying gives the quadratic equation $p^2 - 10p - 1200 = 0$. Solving for p, we find that the meaningful root is $p = 40$ cents per dozen. Thus the new price is 30 cents per dozen. Before the price change, Stuart bought 15 apples; after the price change, he bought 20 apples.

2.11 The discrepancy between the clocks increases by ten minutes each hour. After six hours, the discrepancy is 60 minutes, or one hour.

2.13 There are four different kinds of days: RR, RS, SR, SS, where RS means rain in the morning and sun in the afternoon, and so forth (see table). The given information implies that there are no days on which it rained in the morning and afternoon.

	Rain PM	Sun PM
Rain AM	0	n
Sun AM	k	m

We let k, m, and n denote the other three kinds of days, as shown. Because there were 11 sunny mornings, we have $k + m = 11$. Because there were 12 sunny afternoons, we have $n + m = 12$. Because there were 13 rainy days, we have $k + n = 13$. Solving these equations, we have $k = 6$, $m = 5$, $n = 7$, for a total of 18 days.

2.15 Let t be the number of hours before noon that the sun rises, L be the distance between the towns, x be the distance Pat rides before noon, v_p be Pat's riding speed, and v_d be Dana's riding speed. For the first part of the trip (before noon), we have $tv_p = x$ and $tv_p = L - x$. For the second part of the trip (after noon), we have $5v_p = L - x$ and $8v_d = x$. These equations can be solved to find that $t = 2\sqrt{10}$ hours. Thus, the sun rose $2\sqrt{10} \doteq 6.32$ hours before noon.

2.17(i) The conditions imply that $A = (1 + x)B$ and $B = (1 - y)A$. Eliminating A and B leaves $y = \frac{x}{1+x}$, where x can be any nonnegative number.

2.17(iii) Your age is x and your brother's age is $x + 2$. Thus, your age is $y = \frac{x}{x+2}$ of your brother's age, where $x \geq 0$.

2.17(v) Letting the length of the trip be L, the expected time of travel is $\frac{L}{60}$ hours. Traveling $\frac{L}{2}$ miles (halfway to the destination) at x miles per hour takes $\frac{L/2}{x}$ hours and leaves $\frac{L}{60} - \frac{L/2}{x}$ hours for the second half of the trip. Therefore, the speed required to arrive on time is

$$y = \frac{L/2}{\frac{L}{60} - \frac{L/2}{x}} = \frac{30x}{x - 30}.$$

The domain is $\{x : x > 30\}$, meaning that if the speed for the first half of the trip is less than or equal to 30 miles per hour, it is impossible to arrive on time.

2.17(vii) The number of gallons required is $\frac{x}{32}$, so the cost of the gasoline is $y = \frac{x}{32} \cdot \$1.60 = 0.05x$. The function is defined and makes sense for $x \geq 0$.

2.17(ix) Let L be the total distance traveled by each of A and B when B overtakes A. In time y, A travels a distance $L - 30x$ miles at a speed of 30 miles per hour, and B travels a distance L miles at a speed of 40 miles per hour. Thus, $y = \frac{L-30x}{30} = \frac{L}{40}$. Eliminating L and solving for y, we have $y = 3x$, where $x \geq 0$.

2.17(xi) Let L be the total distance traveled by each of A and B when B overtakes A. In 3 hours, A travels a distance $L - 0.6x$ miles at speed x miles per hour, and B travels a distance L miles at a speed of y miles per hour. Thus, $3 = \frac{L-0.6x}{x} = \frac{L}{y}$. Eliminating L and solving for y, we have $y = 1.2x$. The function is defined and gives nonnegative values of y for $x \geq 0$.

2.18(i) In general, if v liters are removed from V liters of a p percent solution and replaced by v liters of a q percent solution, the resulting solution has a percentage concentration of

$$p_{final} = \frac{(V - v)\frac{p}{100} + v\frac{q}{100}}{V},$$

where p and q are percentages between 0 and 100. In this case, we have

$$0.65 = \frac{(10 - x) \cdot 0.4 + x \cdot \frac{y}{100}}{10}.$$

Solving for y gives the function $y = f(x) = 40 + \frac{250}{x}$. Note that $f(10) = 65$ and $f\left(\frac{25}{6}\right) = 100$, and thus the admissible values of x are $\left\{x : \frac{25}{6} \leq x \leq 10\right\}$. This says that the dilution cannot be done if fewer than $\frac{25}{6}$ liters of solution are removed.

2.18(iii) By part (i), in this case, we have

$$0.5 = \frac{(10 - 2) \cdot \frac{x}{100} + 2 \cdot \frac{y}{100}}{10}.$$

Solving for y gives the function $y = f(x) = 250 - 4x$. Note that $f\left(\frac{125}{2}\right) = 0$ and $f\left(\frac{75}{2}\right) = 100$, and thus the admissible values of x are $\left\{x : \frac{75}{2} \leq x \leq \frac{125}{2}\right\}$. Solutions with concentrations between 0% and 100% can be produced by this dilution.

2.19 Here are the critical observations, where A, B, and C mean algebra, biology, and chemistry, respectively:

- The 2 students who failed at least AB include the 1 student who failed ABC; so 1 student failed only AB.
- The 6 students who failed at least AC include the 1 student who failed ABC; so 5 students failed only AC.
- The 3 students who failed at least BC include the 1 student who failed ABC; so 2 students failed only BC.
- Now the 12 students who failed at least A include the 1 student who failed ABC, the 1 student who failed only AB, and the 5 students who failed only AC; so $12 - 1 - 5 - 1 = 5$ students failed only A.
- The 5 students who failed at least B include the 1 student who failed ABC, the 1 student who failed only AB, and the 2 students who failed only BC; so $5 - 1 - 1 - 2 = 1$ student failed only B.
- Finally, the 5 students who failed at least C include the 1 student who failed ABC, the 5 students who failed only AC, and the 2 students who failed only BC; so $8 - 1 - 5 - 2 = 0$ students failed only C.

We see that six students failed exactly one of the exams, eight students failed exactly two of the exams, and one student failed all three exams. Thus $41 - 6 - 8 - 1 = 26$ students passed all three exams.

2.21 Let N be the number of coconuts in the pile at the beginning. Let S_i be the number of coconuts taken by the ith sailor, and let P_i be the number of coconuts left in the pile after the ith sailor takes his share. Then we have the following relations:

$$S_1 = \frac{1}{5}(N-1) \text{ and } P_1 = \frac{4}{5}(N-1) \quad \text{Note: } S_1 + P_1 = N - 1.$$

$$S_2 = \frac{1}{5}(P_1 - 1) \text{ and } P_2 = \frac{4}{5}(P_1 - 1) \quad \text{Note: } S_2 + P_2 = P_1 - 1.$$

$$S_3 = \frac{1}{5}(P_2 - 1) \text{ and } P_3 = \frac{4}{5}(P_2 - 1) \quad \text{Note: } S_3 + P_3 = P_2 - 1.$$

$$S_4 = \frac{1}{5}(P_3 - 1) \text{ and } P_4 = \frac{4}{5}(P_3 - 1) \quad \text{Note: } S_4 + P_4 = P_3 - 1.$$

$$S_5 = \frac{1}{5}(P_4 - 1) \text{ and } P_5 = \frac{4}{5}(P_4 - 1) \quad \text{Note: } S_5 + P_5 = P_4 - 1.$$

Because the number of coconuts in the final pile was divisible by 5, we have $P_5 = 5m$, where m is a positive integer. Working backwards through the above relations yields

$$N = (5m+4)\frac{5^5}{4^5} - 4.$$

Because N is a positive integer, $5m + 4$ must be divisible by $4^5 = 1024$. The least values of m and N for which these conditions hold are $m = 204$ and $N = 3121$. Clearly, larger values of N are possible.

2.23 [40] A bit of trial and error shows that Ann cannot be assured of selecting *more* than 40 pairs of shoes. For example, the distribution shown in the first table below gives exactly 40 pairs of black shoes and no pairs of brown or white shoes.

	Black	Brown	White	Total
Left	40	80	0	120
Right	40	0	80	120
Total	80	80	80	240

	Black	Brown	White	Total
Left	a	b	c	120
Right	d	e	f	120
Total	80	80	80	240

	Black	Brown	White	Total
Left	0	0	c	120
Right	d	e	0	120
Total	80	80	80	240

However, can Ann always find at least 40 pairs of shoes? To minimize the number of pairs, an indirect approach seems easiest. Suppose that the number of pairs of shoes that Ann is assured of finding is $P < 40$. After Ann has removed these P pairs from the box, suppose that the number of shoes in each category is given by the second table above. At least one number in each column must be zero (if not, then there would be another pair that Ann could remove). However, both numbers in the same column cannot be zero because the total number of shoes of each color (in each column) is 80 and Ann removed only $P < 40$ pairs. Similarly, no row of the table can have all zeros because the number of shoes in each row is 120 and Ann removed only $P < 40$ pairs. Thus, two of the zeros must be in one row, and one of the zeros must be in the other row, as shown in the last table above. It follows that $c = 120 - P$ because P pairs of shoes have been removed from this row. But we also have $c \leq 80$. Taken together, $120 - P \leq 80$ implies that $P \geq 40$. From the initial assumption, we know $P < 40$, so Ann is assured of finding no more than or no less than 40 pairs of shoes.

Chapter 3

3.1 Let s, r, and e denote the number of student, regular, and senior tickets, respectively, sold. Then we have $10s + 20r + 15e = \$125$. Solving for s, the solutions must satisfy

$$s = \frac{125 - 20r - 15e}{10},$$

which means that $125 - 20r - 15e$ must be a multiple of 10, which in turn implies that e must be odd. Solutions may now be enumerated:

r	e	s
0	$\{1, 3, 5, 7\}$	$\{11, 8, 5, 2\}$
1	$\{1, 3, 5, 7\}$	$\{9, 6, 3, 1\}$
2	$\{1, 3, 5\}$	$\{7, 4, 1\}$
3	$\{1, 3\}$	$\{5, 2\}$
4	$\{1, 3\}$	$\{3, 0\}$
5	1	1

3.3 Assume that the woman begins with $0. After the first purchase she is $500 in debt. After the first sale she has $100. After the second purchase she is $600 in debt. After the second sale she has $200. She gained $200 on the transactions.

3.5 Cut the cylinder lengthwise and lay it out flat. The wire now forms the hypotenuse of eight identical right triangles, whose side lengths are $20/8 = 2.5$ centimeters and 6 centimeters. The length of wire is $8\sqrt{2.5^2 + 6^2} = 52$ centimeters.

3.7 The key is to note that as you look at the volumes on the shelf, the first page of the first volume is not on the far left side of the volumes. Similarly, the last page of the last volume is not on the far right side of the volumes. With n volumes, the worm travels $\frac{1}{4} + (n-2) \cdot (1 + \frac{1}{4} + \frac{1}{4}) + \frac{1}{4} = \frac{3}{2}n - \frac{5}{2}$ inch, which reduces to $\frac{1}{2}$ inch for two volumes.

3.9 The circumference is related linearly to the radius by $C = 2\pi r$. Therefore, the change in C is related to the change in r by $\Delta C = 2\pi\Delta r$. We are given that $\Delta C = 2\pi$; therefore, $\Delta r = 1$ foot. Thus the new string is one foot from the Earth everywhere.

3.11 The given numbers imply that 45 nurses play softball only and 25 nurses play soccer only. Taken with the 50 nurses who play both sports, at least 120 nurses must have been surveyed. Thus the survey is flawed.

3.13 We are told that ten buyers bought fewer than six cars and one buyer bought more than nine cars. Therefore, $30 - 10 - 1 = 19$ buyers bought 6, 7, 8, or 9 cars.

3.15 The table shows all sets of ages with a product of 36. The fact that the sum of the ages equals today's date provides two clues. First, the age combination (1, 1, 36) can be eliminated because a date cannot be greater than 31. Second, notice that all the sums are different except those for (1, 6, 6) and (2, 2, 9). So if the ages were any of the other combinations, Paul would be able to figure out the ages. Because he can't figure out the ages, only the combinations (1, 6, 6) and (2, 2, 9) remain. The last clue, "the oldest has red hair," says that there is an oldest child, so (1, 6, 6) is eliminated. The children have ages 2, 2, and 9.

Ages	Sum
(1,1,36)	38
(1,2,18)	21
(1,3,12)	16
(1,4,9)	14
(1,6,6)	13
(2,2,9)	13
(2,3,6)	11
(3,3,4)	10

3.17 Because the miniature statues are one-sixth as tall as the original statue, one miniature statue requires $\left(\frac{1}{6}\right)^2$ as much paint as the original statue, or $\frac{1}{36}$ pints. Thus 540 miniature statues require $\frac{540}{36} = 15$ pints of paint.

3.19 Let s and b be the current ages of the ship and boiler, respectively. The difference in ages is always $d = s - b$. The "age of the boiler when the ship was as old as the boiler is now" is $b - d$. Thus the statement in the problem means $s = 2(b - d)$. Substituting $d = s - b$, we have $3s = 4b$, or $\frac{s}{b} = \frac{4}{3}$. The ratio of the ship's age to the boiler's age is $\frac{4}{3}$.

3.21 As shown in the table, there are four different kinds of round trips. The given information implies that there are no days on which the woman took the train in both the morning and afternoon; thus, $a = 0$. Because the woman took the bus in the morning eight times, we have $c + d = 8$. Because she took the bus in the afternoon 15 times, we have $b + d = 15$. Because she took the train nine times, we have $b + c = 9$. Solving these equations, we have $b = 8$, $c = 1$, $d = 7$, for a total of 16 days.

	Train PM	Bus PM
Train AM	a	b
Bus AM	c	d

3.23 In order to determine a winner, all but one person must be defeated once. Therefore, $N - 1$ games are needed to determine a winner.

3.25 Let p, n, d, q, and h denote the number of pennies, nickels, dimes, quarters, and half-dollars, which must all be nonnegative integers. Then $p + n + d + q + h = 50$ (because there are 50 coins) and $p + 5n + 10d + 25q + 50h = 100$ (because the value of the coins is 100 cents). First note that there is no way to meet the conditions if $h \geq 1$; thus, $h = 0$. With this observation, p can be eliminated from the two equations to give $4n + 9d + 24q = 50$. The easiest way to proceed is to let $q = 0, 1, 2, \ldots$ successively. With $q = 0$, we have $n = 8, d = 2, p = 40$. With $q = 1$, we have $n = 2, d = 2, p = 45$. With $q \geq 2$, there are no solutions. Thus at most two people could have attended the meeting.

3.27 Assume that N people pass through the security system. Of the $\frac{N}{2}$ travelers who are inspected at the first stage, $\frac{N}{8}$ are inspected at the second stage and $\frac{3N}{8}$ are passed at the second stage. Of the $\frac{N}{2}$ travelers who are passed at the first stage, $\frac{N}{8}$ are inspected at the second stage and $\frac{3N}{8}$ are passed at the second stage. This implies that $\frac{N}{8}$ travelers are inspected twice, $\frac{N}{2}$ travelers are inspected once, and $\frac{3N}{8}$ travelers are not inspected.

3.29 Assume that one of the flasks has been set aside. Among the remaining flasks, the amount of wine is twice the amount of water. This implies that the total volume of liquid (water and wine) in the remaining flasks is three times the amount of water used; therefore, the sum of the capacities of the remaining flasks is divisible by 3. There is only one set of five of the six flasks whose total volume is divisible by 3; it is the set $\{16, 18, 22, 24, 34\}$, whose sum is 114. Because $\frac{114}{3} = 38$, the amount of water used is 38 ounces and the amount of wine used is $2 \cdot 38 = 76$ ounces. Thus the 23-ounce flask was not filled, the 16- and 22-ounce flasks were filled with water, and the 18-, 24-, and 34-ounce flasks were filled with wine.

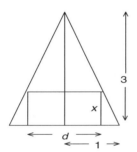

3.31 Imagine slicing the cone along its vertical axis through a diagonal of the base of the cube; the resulting side view is shown in the figure above. Letting x denote the side length of the cube and d the length of the diagonal of the base of the cube, we have $d = \sqrt{2}x$. Using similar triangles in this figure, we see that

$$\frac{1}{3} = \frac{1 - \frac{d}{2}}{x} = \frac{1 - \frac{\sqrt{2}x}{2}}{x}.$$

The solution to this equation is $x = \frac{6}{2+3\sqrt{2}}$, so the volume of the cube is $x^3 \doteq 0.888$.

3.33 The devious solution is to assume that because the radii of the sphere and hole are not given, the volume of the remaining material is independent of them. If this is the case, then consider a sphere of diameter 10 inches, the length of the hole. Then the hole has a volume of zero and the volume of the remaining material is the volume of the entire sphere, or $\frac{500\pi}{3}$ cubic inches.

   Alternatively, assume that the volume of the sphere is r inches. Then the volume of the removed material is the volume of a cylinder ($V_{cyl} = 10\pi(r^2 - 25)$) plus the volume of two spherical caps. The volume of one spherical cap is $V_{cap} = \frac{\pi}{3}h^2(3r - h) = \frac{\pi}{3}(r - 5)^2(3r - (r - 5))$, where $h = r - 5$ is the height of the cap. Thus the volume of the material remaining is $V = \frac{4}{3}\pi r^3 - V_{cyl} - 2V_{cap}$. A bit of algebra reveals that the terms involving r disappear, and thus again $V = \frac{500\pi}{3}$ cubic inches.

3.35 Here are some preliminary observations. The constraints imply that no one shook more than eight hands, so the nine numbers reported to Mr. Schmidt must have been $0, 1, 2, \ldots, 8$. Mr. Schmidt was not included in the poll, so he shook the same number of hands as someone else. Here is the key question: Given that someone shook eight hands (all but him/herself and his/her spouse), how could anyone shake zero hands? It follows that the only person who could shake zero hands is the spouse of the person who shook eight hands. Removing the couple that shook zero and eight hands, it follows successively that another couple shook one and seven hands, another couple shook two and six hands, another couple shook three and five hands, and another couple shook four and four hands. Because only Mr. Schmidt shook the same number of hands as someone else, Mr. and Mrs. Schmidt each shook four hands.

Chapter 4

4.1(ii) The table shows one possible itinerary (H_i and B_i denote the respective husbands and brides). Nine crossings are needed.

Near bank

H_1, H_2, H_3, B_3	H_1, H_2, H_3, B_3	H_1, H_3, B_3	H_1, H_3, B_3	H_1, H_2
↓ B_1, B_2	↑ B_2	↓ H_2, B_2	↑ H_2	↓ H_3, B_3
	B_1	B_1	B_1, B_2	B_1, B_2

Far bank

Near bank

H_1, H_2	H_3	H_3	
↑ H_3	↓ H_1, H_2	↑ H_2	↓ H_2, H_3
B_1, B_2, B_3	B_1, B_2, B_3	H_1, B_1, B_2, B_3	H_1, B_1, B_2, B_3

Far bank

4.2(i) Proceeding algebraically, let p and q be the net number of fillings of the 4- and 7-gallon jugs, respectively, from the well. Then p and q must satisfy $4p + 7q = 2$. One solution is $p = -3$ and $q = 2$. Thus the 4-gallon jug must be emptied 3 times and the 7-gallon jug must be filled twice. The sequence of transfers is shown below, where (a, b) means the 4-gallon jug contains a gallons and the 7-gallon jug contains b gallons.

Jug status	Move
(0,0)	Both jugs empty
(0,7)	Fill 7-gallon jug from well
(4,3)	Empty 7-gallon jug into 4-gallon jug
(0,3)	Empty 4-gallon jug into well
(3,0)	Empty 7-gallon into 4-gallon jug
(3,7)	Fill 7-gallon jug from well
(4,6)	Empty 7-gallon jug into 4-gallon jug
(0,6)	Empty 4-gallon jug into well
(4,2)	Empty 7-gallon jug into 4-gallon jug
(0,2)	Empty 4-gallon jug into well

Other longer sequences are possible.

4.2(iii) Proceeding algebraically, let p and q be the net number of fillings of the 7- and 11-gallon jugs, respectively, from the well. Then p and q must satisfy $7p+11q = 2$. One solution is $p = 5$ and $q = -3$. Thus the 7-gallon jug must be filled 5 times and the 11-gallon jug must be emptied 3 times. The sequence of transfers is shown below, where (a, b) means that the 7-gallon jug contains a gallons and the 11-gallon jug contains b gallons.

Jug status	Move
(0,0)	Both jugs empty
(7,0)	Fill 7-gallon jug from well
(0,7)	Empty 7-gallon jug into 11-gallon jug
(7,7)	Fill 7-gallon jug from well
(3,11)	Empty 7-gallon jug into 11-gallon jug
(3,0)	Empty 11-gallon jug into well
(0,3)	Empty 7-gallon jug into 11-gallon jug
(7,3)	Fill 7-gallon jug from well
(0,10)	Empty 7-gallon jug into 11-gallon jug
(7,10)	Fill 7-gallon jug from well
(6,11)	Empty 7-gallon jug into 11-gallon jug
(6,0)	Empty 11-gallon jug into well
(0,6)	Empty 7-gallon jug into 11-gallon jug
(7,6)	Fill 7-gallon jug from well
(2,13)	Empty 7-gallon jug into 11-gallon jug
(2,0)	Empty 11-gallon jug into well

4.2(v) Let (a/b) indicate that an hourglass has a minutes in the upper chamber and b minutes in the lower chamber. The following table shows one way to measure nine minutes.

4-minute glass	7-minute glass	Time elapsed
(4/0)	(7/0)	0 minutes
(0/4)	(3/4)	4 minutes
(4/0)	(3/4)	4 minutes
(1/0)	(0/7)	7 minutes: START NOW
(0/1)	(0/7)	8 minutes
(4/0)	(0/7)	8 minutes
(0/4)	(0/7)	12 minutes
(4/0)	(0/7)	12 minutes
(0/4)	(0/7)	16 minutes

Using this scheme, 9 minutes elapse after the 7-minute start-up period; so the entire process takes 16 minutes.

4.3(i) Fill the cylinder to the top, then slowly pour the water from the cylinder until the water level forms a horizontal line between the lip of the cylinder and the highest point on the base of the cylinder. In this way, the rectangular cross section of the cylinder is bisected diagonally by the water line.

4.3(ii) The key is to realize that the whiskey in the 10-quart cylinder can be halved using the tilting process described in part (i). Other than that, some trial and error is needed. Here is a sequence of transfers that seems as short as possible. We let (a, b, c, d) denote the contents of the 10-, 11-, and 13-quart cylinders and the 24-quart barrel.

Container status	Move
(0,0,0,24)	Cylinders empty
(10,11,0,3)	Fill 10-qt and 11-qt from barrel
(0,11,10,3)	Empty 10-qt into 13-qt
(0,8,13,3)	Pour 11-qt into 13-qt
(10,8,3,3)	Pour 13-qt into 10-qt
(5,8,8,3)	Halve 10-qt and pour into 13-qt
(8,8,8,3)	Empty barrel into 10-qt

4.4(i) Place three coins in each pan. If the pans balance, the light counterfeit is one of the two remaining coins, and can be found in one more weighing. If the pans do not balance, the light counterfeit is in the pan that rises, and can be found in one more weighing.

4.4(iii) Divide the coins into groups of 27, 27, and 26, weighing the two groups of 27 against each other. If the pans balance, the (light) counterfeit is among the 26 coins. If the pans do not balance, the (light) counterfeit is in the pan that rises. Either way, the counterfeit is located in a pile of at most 27 coins. Two more weighings place the counterfeit in a group of nine coins and then a group of three coins. The last weighing is used to identify the counterfeit among a group of three coins.

4.4(v) Denote the coins $C_1, \ldots, C_5$. Choose two coins, C_1 and C_2, and weigh them against each other. If the pans balance, the counterfeits are among the remaining three coins, all of which have different weights. The light and heavy counterfeits can be identified in two weighings by weighing these coins against a good coin. If the pans do not balance in the first weighing, assume without loss of generality that the left pan with C_1 drops and the right pan with C_2 rises. Then either C_1 is heavy and/or C_2 is light, but both cannot be good. For the second and third weighings, it is important to note the movement of the pans relative to their previous position.

For the second weighing, add C_3 to the left pan (with C_1) and add C_4 to the right pan (with C_2). Nine cases follow; all of them use a third weighing with C_1 and C_3 on the left pan (same as the second weighing) and with C_2 and C_5 on the right pan (replacing C_4 by C_5 after the second weighing).

If the pans balance on the second weighing and the pans balance on the third weighing, C_3, C_4, C_5 are good, C_1 is heavy, and C_2 is light. If the pans balance on the second weighing and the left pan drops on the third weighing, C_2, C_3, C_4 are good, C_1 is heavy, and C_5 is light. If the pans balance on the second weighing and the left pan rises on the third weighing, C_1, C_3, C_4 are good, C_5 is heavy, and C_2 is light.

If the left pan drops on the second weighing and the pans balance on the third weighing, C_1, C_4, C_5 are good, C_2 is light, and C_5 is heavy. If the left pan drops on the second weighing and the left pan drops on the third weighing, then the left pan has dropped three times, which is impossible. If the left pan drops on the second weighing and the left pan rises on the third weighing, C_2, C_3, C_5 are good, C_4 is light, and C_1 is heavy. In all cases, the counterfeits can be identified.

If the left pan rises on the second weighing and the pans balance on the third weighing, C_2, C_4, C_5 are good, C_3 is light, and C_1 is heavy. If the left pan rises on the second weighing and the left pan drops on the third weighing, C_1, C_3, C_5 are good, C_2 is light, and C_4 is heavy. If the left pan rises on the second weighing and the left pan rises on the third weighing, then we again have an impossible situation.

4.4(vii) Place four coins in each pan. If the pans balance, only four coins remain to be tested. The counterfeit can be found by weighing two of the four remaining coins against two good coins. If the pans do not balance on the first weighing, we have four good coins (G), four possible light coins (L), and four possible heavy coins (H). For the second weighing, place 3H and 1L on the left pan and place 1H and 3G on the right pan; this leaves 3L unused. If the pans balance, the counterfeit is among the 3L, and can be found in one more weighing. If the left pan drops, the counterfeit is one of the 3H on the left pan, and can be found in one more weighing. If the left pan rises, the counterfeit is the possible L on the left pan or the possible H on the right pan. It can be found in one more weighing.

4.4(ix) Choose two coins as test coins and weigh them against each other. Case 1: If they do not balance, then one is steel and one is brass. Now weigh the two test coins successively against two untested coins. If the two test coins balance the two untested coins, then one of the untested coins is steel and one is brass (remember, we only need to count each type of coin). If the pan with the two test coins rises, then both untested coins are steel. If the pan with the two test coins drops, then both untested coins are brass. In all three cases we can count the number of brass and steel coins in twelve weighings (one for the initial test coins and eleven weighings for the eleven remaining pairs of coins).

Case 2: If the two test coins do balance, either both are steel or both are brass. Now weigh the two test coins successively against two untested coins. As long as the pans balance, continue to count the number of coins that are the same as the test coins (as yet it is not known if the test coins are steel or brass). Eventually, the two test coins will not balance the two untested coins. If the pan with the test coins rises, at least one of the untested coins is a steel coin and all of the coins weighed up to this point are brass. If the pan with the test coins drops, at least one of the untested coins is a brass coin and all of the coins weighed up to this point are steel. In either case, weighing the two untested coins against each other determines whether they are identical or different. Call this weighing the *extra weighing*. It is now possible to choose two new test coins, one steel and one brass. Now proceed as in case 1, weighing the remaining pairs of coins against the two new test coins. The number of weighings is the same as in case 1 plus the *extra weighing*. So 13 weighings are needed.

4.5 Choose three pairs of coins, weigh them, and call their weights w_1, w_2, and w_3. The counterfeit must be in one of these pairs, so without loss of generality, assume it is in the third pair of coins, in which case, $w_1 = w_2$. We now know that the regular coins each weigh $w = \frac{1}{2}w_1 = \frac{1}{2}w_2$. If $w_3 < w_1 = w_2$, then the counterfeit is light; if $w_3 > w_1 = w_2$, then the counterfeit is heavy. In either case, the weight of the counterfeit coin, x, satisfies $w + x = w_3$ or $x = w_3 - w$.

4.7 Turn on one of the switches and leave it on for a short while. Turn off the first switch and turn on a second switch. While the second switch is on, go upstairs and determine which lamp is off and hot (corresponding to the first switch), which lamp is on (corresponding to the second switch), and which lamp is off and cold (corresponding to the third switch).

4.9 Move the top coin to the right end of the second row from the top. Move the lower left corner coin to the left end of the second row. Move the lower right corner coin to form the point below the fourth row.

4.11 Surgeon 1 puts on one pair of gloves (call it the blue pair) and then puts the other pair of gloves (call it the red pair) over the blue pair. We will call the outside of the gloves side 1 and the inside of the gloves side 2. After Surgeon 1 has operated, the outside of the red pair (side 1) and the inside of the blue pair (side 2) are contaminated. Surgeon 2 now puts on the red pair and operates, which is allowed because the inside of the red pair (side 2) is sterile; the outside of the red pair will have touched the patient twice, which is allowed. Now, Surgeon 3 turns the blue gloves inside out and puts them on. Note that the sterile side 1 of the blue gloves now touches the Surgeon 3; however, side 2 of the blue gloves which has touched Surgeon 1 will touch the patient, which is not allowed. Therefore, Surgeon 3 must put the red pair of gloves over the blue pair, which is allowed because the outside of the red pair has touched only the patient.

4.13 The store owner gives the boy a thin box that is 4 feet long and 3 feet wide. The pole fits along the 5-foot diagonal of the box, and the box meets the conditions for the bus.

4.15 Cut the three links of one of the chains. Use one of the cut links to join two of the five chains to make a chain with six links. Use another of the cut links to join two of the five chains to make a chain with six links. Use the third cut link to join the two chains of six links. Then weld the three cut links. Three links have been cut and welded, so the cost is $9.

4.17 Let DWF mean day's worth of fuel. Let FS mean fuel stop.

- Days 1-2: Load three DWF, leave one DWF at FS 1, return to Earth.
- Days 3-4: Load three DWF, leave one DWF at FS 1, return to Earth.
- Days 5-6: Load three DWF, leave one DWF at FS 1, return to Earth (FS 1 now has three DWF).
- Days 7-10: Load three DWF; fly to FS 1, refuel; fly to FS 2, leave one DWF at FS 2; fly to FS 1, refuel; return to Earth (FS 1 and 2 now each have one DWF).
- Days 11-15: Load three DWF; fly to FS 1, refuel; fly to FS 2, refuel; continue to the space station.

4.19 The following scheme requires twelve steps, which seems to be optimal. Note that Q = queen, D = daughter, S = son, R = rock, P = pig, G = dog, C = cat.

Step	0	1	2	3	4	5
Tower	QDSPGC	QDPGCR	QSPGCR	QSPGC	DSPGCR	DSPGC
Ground	R	S	D	DR	Q	QR

Step	6	7	8	9	10	11	12
Tower	DSPR	DSP	SPCR	SPC	SR	S	R
Ground	QGC	QGCR	QDG	QDGR	QDPGC	QDPGCR	QDPGCS

Chapter 5

5.1 Because after-tax income is 27% less than before-tax income, before-tax income is $34,500/0.73 = $47,260.

5.3 Let R be the original retail price. The retail price that Jack paid was $0.75R$. With tax, Jack paid a total of $(0.75R)1.076 = 275.78. Solving, we see that $R = $341.73.

5.5 The price for the perfume is $1800 per ounce for either 1/60 or 1/30 ounce. Letting x be the price for 1/45 ounce, we need $\frac{x}{1/45} = 1800$, or $x = $40. Therefore the price for 1/45 ounce should be $40.

5.7 Let x be the required grade on the final exam. Then

$$\frac{185}{200} \cdot 0.55 + \frac{45}{50} \cdot 0.15 + \frac{81}{100} \cdot 0.15 + x \cdot 0.15 = 0.90.$$

Solving for x, the required final exam grade needed for a 90% average is 89.83%.

5.9 The price after applying the three coupons is $p(0.75)(0.65)(0.60) = 0.29\,p$, where p is the original price. Thus the total discount is 71%.

5.11 Suppose that Sammy had a fraction p of his at-bats in the first half, and a fraction $1 - p$ of his at-bats in the second half. Then his overall average is $0.333\,p + 0.500(1 - p) = 0.386$. Solving, we find that $p = 0.683$. Similarly, suppose that Barry had a fraction q of his at-bats in the first half, and a fraction $1 - q$ of his at-bats in the second half. Then his overall average is $0.200q + 0.467(1 - q) = 0.400$. Solving, we find that $q = 0.251$. The apparent paradox is explained by the fact that Sammy had nearly two-thirds of his at-bats in the first half, when he had a relatively low average, while Barry had three-fourths of his at-bats in the second half, when he had a relatively high average.

5.13 Let Q_1 and Q_2 be the July and August sales, respectively. Let I_1 and I_2 be the July and August incomes, respectively. Let p be the August price of chocolate per pound. Then we have $Q_2 = 0.768Q_1$, $I_2 = 0.876I_1$, $I_1 = \$2.56Q_1$, and $I_2 = pQ_2$. Combining these relations, we find that

$$p = \frac{\$2.56 \cdot 0.876}{0.768} = \$2.92 \quad \text{per pound.}$$

5.15 Let the number of men and women be M and W, respectively. By the conditions of the problem, we have $\frac{M}{W} = \frac{17}{15}$ and $\frac{M-90}{W-80} = \frac{8}{7}$. Solving gives $M = 170$ and $W = 150$.

5.17(i) Using the incidence rate, of the 4000 people tested, 1.5%, or 60 people, have the disease; therefore, 3940 do not have the disease. Of the 60 people with the disease, 80%, or 48 people, test positive (true positives), and 12 people test negative (false negatives). Of the 60 people without the disease, 80%, or 3152 people, test negative (true negatives), and 788 people test positive (false positives).

5.17(ii) Of the 60 people with the disease, 48 test positive, so the probability that a person with the disease tests positive is $\frac{48}{60} = 0.8$.

5.17(iii) Of the 836 people who test positive, 48 have the disease, so the probability that a person who tests positive has the disease is $\frac{48}{836} = 0.057$.

5.17(iv) A patient who has tested positive should be told the probability that, having tested positive, s/he has the disease. (This probability is 0.057.)

5.19(i) The following table helps organize the solution:

	American	Swedish	Other	Total
Men	a	b	4	$N - 24$
Women	c	d	4	24
Totals	15	$N - 23$	8	N

We see that $a + c = 15$ and $c + d = 20$. Combining these relations, we have $d = a + 5$, which means there are five more Swedish women than American men.

5.19(ii) The first row of the above table implies that $a + b = N - 28$, where N is the total number of employees. If $a \geq 1$, then this fact implies that $N \geq 29 + b$. Nothing in the problem precludes $b = 0$; thus, $N \geq 29$.

5.21 Assume that each jar contains v (volume) units of solution. The amount of alcohol in the first jar is $\frac{pv}{p+1}$; the amount of water in the first jar is $\frac{v}{p+1}$; the amount of alcohol in the second jar is $\frac{qv}{q+1}$; and the amount of water in the second jar is $\frac{v}{q+1}$. The ratio of total alcohol to total water in the combined mixture is

$$\frac{\frac{pv}{p+1} + \frac{qv}{q+1}}{\frac{v}{p+1} + \frac{v}{q+1}} = \frac{p + q + 2pq}{p + q + 2}.$$

5.23 The fees increased by a factor of 1.07 and the number of students increased by a factor of 1.01. The fraction of the fees paid personally decreased by $\frac{0.03}{0.6} = 0.05$ to 0.95 of its original value. Thus the fees collected directly from students changed to $1.07 \cdot 1.01 \cdot 0.95 = 1.026$ of its original value, which is an increase of 2.6%. Hat derivatives would give a change of $7\% + 1\% - 5\% = 3\%$.

5.25 With maximum overlap between the groups, the 90 who arrived by bus include the 75 women, who include the 65 from Alaska, who include the 45 with gloves. Thus, at most 45 people fit the description.

To find the minimum overlap, note that at a minimum, 65 people are women arriving by bus; at a minimum, 35 people are women from Alaska arriving by bus. This means that the 45 people with gloves can be distributed among the 65 people who are not women from Alaska arriving by bus. Thus at a minimum, there are no people who meet the description.

5.27 Let a_1 and a_2 be the number of patients receiving Treatment A in the first and second trials, respectively. Let b_1 and b_2 be the number of patients receiving Treatment B in the first and second trials, respectively. The overall cure rate for Treatment A implies

$$\frac{0.2a_1 + 0.85a_2}{a_1 + a_2} = 0.417.$$

The overall cure rate for Treatment B implies

$$\frac{0.15b_1 + 0.75b_2}{b_1 + b_2} = 0.55.$$

The first equation implies that $a_1 = \frac{0.433}{0.217}a_2 = 2a_2$. The second equation has a solution, provided that $b_2 = 2b_1$. Twice as many people receive Treatment A in the first trial as in the second trial. Twice as many people receive Treatment B in the second trial as in the first.

5.29 Let M and W be the number of men and women in town, respectively. The number of married men is $0.3M$ and the number of married women is $0.4W$. Because every man and woman has one spouse, we have $0.3M = 0.4W$, or $W = \frac{3}{4}M$. The fraction of townspeople who are unmarried men is $\frac{0.7M}{M+W} = \frac{0.7M}{1.75M} = 0.4$.

5.31 Let the batting average with n at-bats be $0 < a < 1$. Then $a_h = \frac{an+1}{n+1}$ is the new average with a hit, and $a_n = \frac{an}{n+1}$ is the new average with no hit. The amount that the average increases with a hit is $D_h = a_h - a = \frac{1-a}{n+1}$. The amount the average decreases with no hit is $D_n = a - a_n = \frac{a}{n+1}$. We see that if $a > 0.5$, then $D_h < D_n$, and if $a < 0.5$, then $D_h > D_n$. With an average of $a = 0.350$, the average increases more with a hit than it decreases without a hit. The conclusion depends on the current average, but not on the number of hits.

Chapter 6

6.1 Let d miles be the distance traveled by the car either up or down the hill. It spent $\frac{d}{10}$ hours going up the hill and $\frac{d}{20}$ hours going down the hill.Thus, the average speed for the round trip was

$$\frac{2d}{\frac{d}{10} + \frac{d}{20}} = \frac{40}{3} \text{ miles per hour.}$$

6.3 The head of the train must travel 2 miles for the entire train to pass through the tunnel. At 15 miles per hour, it will take $\frac{2}{15}$ hours for the train to pass through the tunnel.

6.5 Let v be the speed of the boat relative to the water and let c be the speed of the current, in units of miles per minute. The downstream trip implies that $\frac{10}{v+c} = 9$. The upstream trip implies that $\frac{10}{v-c} = 15$. These two equations can be solved for v and c to give $v = \frac{8}{9}$ and $c = \frac{2}{9}$ miles per minute. The time required to travel 10 miles with no current is $\frac{10}{8/9} = \frac{45}{4}$ minutes.

6.7(i) Let d be the distance between Denver and Omaha. Let x be the distance between Denver and the passing point. The Denver train reached the passing point in $\frac{x}{80}$ hours, which equals the time traveled by the Omaha train to the passing point, which was $\frac{d-x}{100}$. Equating these times, we find that $\frac{x}{d} = \frac{4}{9}$. The Denver train completed $\frac{4}{9}$ of its trip.

6.7(iii) Let $d = 500$ be the distance between Denver and Omaha. Let x be the distance between Denver and the passing point. The Denver train reached the passing point in $\frac{x}{80}$ hours. The Omaha train reached the passing point in $\frac{500-x}{100}$ hours, which is two hours less than the time traveled by the Denver train. Thus, $\frac{x}{80} - 2 = \frac{500-x}{100}$. Solving for x, the Denver train had traveled $\frac{2800}{9} = 311.1$ miles when the trains passed.

6.7(v) In the hour before the trains met, the New York train traveled 60 miles and the Boston train traveled 70 miles. So the trains were $60 + 70 = 130$ miles apart one hour before they met.

6.9(i) Let a, b, and c be the work rates of Arlen, Ben, and Carla, respectively, in units of jobs per hour. Then $b = \frac{1}{2.5}$ jobs per hour and $c = \frac{1}{3.5}$ jobs per hour. The work rate when all three people work together is $a + b + c$. Thus, $\frac{1}{a+b+c} = 1$ hour. Substituting for b and c and solving for a, we see that Arlen's work rate is $a = \frac{11}{35}$ jobs per hour. Thus, Arlen can do the job in $\frac{35}{11} = 3.2$ hours working alone.

6.9(iii) The work rate is $r = \frac{4 \text{ hats}}{20 \text{ people} \cdot 2 \text{ hours}} = \frac{1}{10}$ hats/person-hours. The time required for 15 people to make 30 hats is

$$\frac{30 \text{ hats}}{r \cdot 15 \text{ people}} = 20 \text{ hours.}$$

The number of people needed to make 40 hats in 4 hours is

$$\frac{40 \text{ hats}}{r \cdot 4 \text{ hours}} = 100 \text{ people.}$$

The number of hats that can be made by 5 people in 12 hours is $r \cdot 5 \text{ people} \cdot 12 \text{ hours} = 6$ hats.

6.9(v) Let a, b, and c be the work rates of Ann, Betty, and Carol, respectively, in units of jobs per days. The three conditions can be written $\frac{1}{a+b} = 10$, $\frac{1}{a+c} = 12$, and $\frac{1}{b+c} = 20$. Solving, we find that $a = \frac{1}{15}$, $b = \frac{1}{30}$, and $c = \frac{1}{60}$ jobs per day. Thus, it takes Carol 60 days to do the job alone.

6.11 The fill rate of Pipe A is $a = \frac{1}{2}$ tanks per hour. The fill rate of Pipe B is $b = \frac{1}{3}$ tanks per hour. The empty rate of Tank C is $c = \frac{1}{5}$ tanks per hour. The net fill rate of the three pipes is $a + b - c = \frac{19}{30}$ tanks per hour. Thus, the time to fill the tank is $\frac{30}{19} \doteq 1.58$ hours.

6.13 The fill rates of the pipes are $a = \frac{1}{10}$ tanks per hour and $b = \frac{1}{8}$ tanks per hour. The empty rate of the drain is $d = \frac{1}{6}$ tanks per hour. The time to fill the first half of the tank is $\frac{1}{2}/\left(\frac{1}{10} + \frac{1}{8} - \frac{1}{6}\right) = \frac{60}{7}$ hours. The time to fill the second half of the tank is $\frac{1}{2}/\left(\frac{1}{10} + \frac{1}{8}\right) = \frac{20}{9}$ hours. The total time to fill the tank is $\frac{60}{7} + \frac{20}{9} \doteq 10.8$ hours.

6.15 Let d be the length of the race. The slower cyclist finishes the race in $\frac{d}{25}$ hours, and the faster cyclist finishes the race in $\frac{d}{30}$ hours. Because the difference in the finishing times is one hour, we have $\frac{d}{25} - \frac{d}{30} = 1$. Solving for d, the length of the race is 150 miles.

6.17 Let $t = 0$ be the time at which the candles are lit. Without loss of generality, assume that the length of each candle is 1. The lengths of the candles are given by $\ell_1(t) = 1 - \frac{t}{6}$ and $\ell_2(t) = 1 - \frac{t}{3}$. The condition that $\ell_1(t) = 2\ell_2(t)$ implies that $t = 2$ hours.

6.19 Let r_A and r_B be the speeds of the boats, where $r_A = \frac{3}{2} r_B$. Boat B traveled a distance $\frac{3}{2} r_B$ during its headstart. Assume Boat B traveled another x miles before it was overtaken. From the moment Boat A started, Boat A traveled $\frac{3}{2} r_B + x$ miles until it overtook Boat B, and Boat B traveled x miles before it was overtaken. Equating the travel times from the moment Boat A started, we have

$$t = \frac{\frac{3}{2} r_B + x}{r_A} = \frac{x}{r_B}.$$

Using $r_A = \frac{3}{2} r_B$, we find that the travel time of Boat A was $t = x/r_B = 3$ hours.

6.21 Let d be the length of a single one-way trip. Let w be the wind speed, and let v_g be the speed of the plane relative to the ground with no wind. Then $u = v_g + w$ is the speed of the plane flying downwind and $v = v_g - w$ is the speed of the plane flying upwind. Combining these two relations, we see that $v_g = \frac{1}{2}(u + v)$. The outbound trip (upwind) implies that $\frac{d}{v} = 84$. The return trip (downwind) implies that $\frac{d}{u} = \frac{d}{v_g} - 9$. Let $t = \frac{d}{u}$ be the time for the return trip. Using all of these relationships, we have

$$t = \frac{d}{u} = \frac{d}{v_g} - 9 = \frac{d}{\frac{1}{2}(u + v)} - 9 = \frac{2}{\frac{1}{t} + \frac{1}{84}} - 9.$$

Simplifying this equation, we find that $t^2 - 75t + 756 = 0$, which has roots $x = 63$ and $x = 12$. The return trip took either 63 or 12 minutes.

6.23 Let r_a and r_z be the rates at which A and Z descend in steps per minute. We know that $r_a = 2r_z$. Let r_e be the rate at which escalator steps appear in steps per minute. Finally, let t_a and t_z be the times needed by A and Z to descend. Then we have

$$r_a = \frac{27}{t_a} = 2r_z = 2\left(\frac{18}{t_z}\right).$$

This relation implies that $t_a = \frac{3}{4} t_z$. The key observation is that the number of steps showing, n, is the number of steps taken by A plus the number of escalator steps that have appeared during the descent; that is $n = 27 + r_e t_a$. Similarly, $n = 18 + r_e t_z$. Using $t_a/t_z = \frac{3}{4}$ and solving for n gives $n = 54$ steps showing.

6.25 Briefly, because the trip took 10 minutes less than usual, the husband drove 5 minutes less than usual to meet his wife and 5 minutes less than usual driving her home. Of her 30-minute headstart, the woman spent 5 fewer minutes driving with her husband. Thus, she spent 25 minutes walking. Alternatively, using some algebra, suppose the husband's usual one-way drive time is t minutes. Then he usually leaves home at 6:00$-t$. This time he left home at 6:00$-t$, as usual, but drove for only $t - 5$ minutes, which means that he met his wife at 5:55. So the wife walked for 25 minutes.

6.27 Let v_p and v_c be the speed of the parade and cyclist, respectively. Let t_1 and t_2 be the time required for the cyclist's forward and backward trips. The goal is to find d, the distance traveled by the cyclist. Because the parade traveled 6 kilometers in the time $t_1 + t_2$, we have $v_p(t_1 + t_2) = 6$. Because the cyclist traveled d kilometers in the time $t_1 + t_2$, we have $v_c(t_1 + t_2) = d$. Combining these two expressions, we see that $d = \frac{6v_c}{v_p}$. So if we can find the ratio of the speeds, we will have d. We now use the fact that the parade was 4 kilometers long. On the forward trip, the speed of the cyclist *relative to the parade* was $v_c - v_p$. Thus, the time required for the forward trip was

$$t_1 = \frac{4}{v_c - v_p}.$$

On the return trip, the speed of the cyclist *relative to the parade* was $v_c + v_p$. Thus, the time required for the return trip was

$$t_2 = \frac{4}{v_c + v_p}.$$

Using t_1 and t_2 in the expression $v_p(t_1 + t_2) = 6$, we find that

$$\frac{4}{v_c - v_p} + \frac{4}{v_c + v_p} = \frac{6}{v_p}.$$

This turns out to be a quadratic equation, $3r^2 - 4r - 3 = 0$, for the ratio $r = \frac{v_c}{v_p}$. The relevant (positive) root is

$$r = \frac{2 + \sqrt{13}}{3}.$$

Thus the distance traveled by the cyclist is

$$d = 6r = \frac{6v_c}{v_p} = 4 + 2\sqrt{13} \doteq 11.21 \text{ kilometers.}$$

6.29 The speed of Bob relative to the truck is 15 km/hr. In 1 minute $= \frac{1}{60}$ hr, Bob travels $\frac{1}{4}$ km relative to the truck. Thus, Bob is $\frac{1}{4}$ km behind the truck one minute before the crash.

6.31 Because Tank B has an outflow rate of 4 gallons per minute and an inflow rate of 3 gallons per minute, the net outflow rate of tank B is 1 gallon per minute. The volume of water in Tank A is given by $V_a = 90 - 3t$, and the volume of water in Tank B is given by $V_b = 90 - t$, where $t \geq 0$. Tank A is empty when $V_a = 0$ or when $t = 30$ minutes. At this time the volume of water in Tank B is $V_b(30) = 60$ gallons. Without the inflow from Tank A, the volume of water in Tank B is now given by $V_b = 60 - 4t$, where $t \geq 30$. Tank B is empty when $V_b = 0$, or $t = 15$ minutes (in addition to the 30 minutes already past). Thus, the time needed to empty both tanks is 45 minutes.

6.33 Let G and g be Alan's and Brenda's harvesting rate in the garden. Let A and a be Alan's and Brenda's harvesting rate in the apple orchard. Then the first condition implies that $\frac{1}{G+g} = 3$ hours and $\frac{1}{g} = 12$. Solving for G, we find that $G = \frac{1}{4}$, which means that working alone, Alan needs 4 hours to harvest the garden. The second condition implies that $\frac{1}{A+a} = 2$ hours and $\frac{1}{A} = 10$. Solving for a, we find that $a = \frac{2}{5}$, which means that working alone, Brenda needs $\frac{5}{2}$ hours to harvest the orchard. Clearly, strategy A requires 5 hours. Using strategy B, Alan and Brenda work independently for $\frac{5}{2}$ hours, and then Brenda joins Alan in the garden, where $\frac{3}{8}$ of the garden remains to be harvested. Working together, they complete harvesting the garden in $3 \cdot \frac{3}{8} = \frac{9}{8}$ hours. Therefore strategy B requires $\frac{5}{2} + \frac{9}{8} = \frac{29}{8}$ hours; thus strategy B is faster.

6.35(i) We have $\theta_p(t) = pt$, $\theta_q(t) = qt$, and $\phi(t) = (p-q)t$. The condition for the runners passing is $\phi(t) = k$, where k is a positive integer. The passing times are $t^* = \frac{k}{p-q}$. At the passing times, the phases are integers, provided that $\theta_p(t^*) = \frac{pk}{p-q} = m$ and $\theta_q(t^*) = \frac{qk}{p-q} = n$, where m and n are positive integers. Thus if $\frac{p}{p-q}$ is an integer, then $\theta_p(t^*)$ is an integer. Because $\frac{p}{p-q} = 1 + \frac{q}{p-q}$, then $\theta_q(t^*)$ is also an integer. This means that the runners pass each other only at the starting point every $\frac{p}{p-q}$ laps of Polly.

6.35(ii) Continuing the above argument, if $\frac{p}{p-q}$ is rational (which implies that $\frac{q}{p-q}$ is rational), then $\theta_p(t^*) = \frac{pk}{p-q}$ and $\theta_q(t^*) = \frac{qk}{p-q}$, where k is an integer, can take on both integer and rational values that differ by an integer. This means that the runners pass each other regularly at both the starting point and other points on the track.

6.35(iii) Finally, if $\frac{p}{p-q}$ is irrational (which implies that $\frac{q}{p-q}$ is irrational), then $\theta_p(t^*) = \frac{pk}{p-q}$ and $\theta_q(t^*) = \frac{qk}{p-q}$, where k is an integer, cannot take on integer values. However, $\theta_p(t^*)$ and $\theta_q(t^*)$ can still differ by an integer. Therefore runners pass each other, but never at the starting point.

6.37 On the first day, Skip ran $\frac{4}{3}$ laps in the time that Harry ran $\frac{1}{3}$ of a lap. Therefore, Skip ran four times faster that Harry. We can (arbitrarily) take Skip's pace to be $p = 4$ laps per minute and Harry's pace to be $q = 1$ lap per minute. On the second day, because they ran in opposite directions, the phase difference was $\phi(t) = (p+q)t = 5t$. The passing times correspond to integer values of the phase difference, or $t^* = \frac{k}{5}$, where k is a positive integer. Harry completed his first lap at $t = 1$, and they passed each other when $t = \frac{1}{5}, \frac{2}{5}, \frac{3}{5}, \frac{4}{5}$. Thus the runners met four times.

6.39 Let v be the speed of the freight train. Let $V = nv$ be the speed of the passenger train. The time required for the trains to pass each other moving in the same direction is $t_1 = \frac{L}{V-v}$, where L is the sum of the lengths of the two trains. The time required for the trains to pass each other moving in the opposite direction is $t_2 = \frac{L}{V+v}$. We are told that $t_1 = nt_2$. Combining these relations, we have $\frac{nL}{(n+1)v} = \frac{L}{(n-1)v}$. Canceling factors of L and v, we are left with a quadratic equation, $n^2 - 2n - 1 = 0$, to be solved for n. The relevant (positive) root is $n = 1 + \sqrt{2}$.

6.41 <u>Solution 1</u>: The speed of the boat relative to the *ground* was $6 - 2 = 4$ km/hr on the upstream trip and $6 + 2 = 8$ km/hr on the downstream trip. So the man paddled for $\frac{3}{4}$ hours upstream before discovering his hat was gone. In that time the hat floated $\frac{3}{2}$ km, so the man was $3 + \frac{3}{2} = \frac{9}{2}$ km from the hat. Let t be the number of hours the hat floated before the man reached it. In the same time, the man traveled $\left(\frac{9}{2} + 2t\right)$ km at a speed of 8 km/hr. Therefore, the man traveled for $\left(\frac{9}{2} + 2t\right)/8$ hours. Thus, $t = \left(\frac{9}{2} + 2t\right)/8$, or $t = \frac{3}{4}$ hours. The man traveled $\frac{3}{4}$ hours upstream and $\frac{3}{4}$ hours downstream, so it was 1:30 when he overtook his hat.

<u>Solution 2</u>: Imagine you are sitting on the hat. From this perspective, the man simply traveled upstream for t hours and downstream for t hours. He traveled a distance $(3+2t)$ km each way at a speed of 6 km/hr, which takes $(3 + 2t)/6$ hours. Thus, $t = (3 + 2t)/6$, or $t = \frac{3}{4}$ hours. The whole trip took $t = \frac{3}{2}$ hours, so it was 1:30 when he overtook his hat.

6.43 Suppose that the maximum volume of grass in the pasture is 1 (one pasture). Let c, h, and s be the rate at which the cow, horse, and sheep graze, respectively, in units of pastures per day. Let g be the rate at which the grass grows, in pastures per day. We are given four pieces of information:

$$\frac{1}{c+h+s-g} = 20,$$

$$\frac{1}{c+h-g} = 25,$$

$$\frac{1}{c+s-g} = \frac{100}{3},$$

$$\frac{1}{h+s-g} = 50.$$

After rearrangement, we have the linear system of equations

$$20c + 20h + 20s - 20g = 1,$$
$$25c + 25h - 25g = 1,$$
$$100c + 100s - 100g = 3,$$
$$50h + 50s - 50g = 1.$$

After some work, the solution $c = 0.03$, $h = 0.02$, $s = 0.01$, $g = 0.01$ emerges. The number of days that the grass will support the cow alone is $\frac{1}{c-g} = 50$ days. Similarly, the grass will support the horse alone for $\frac{1}{h-g} = 100$ days. It will support the sheep for $\frac{1}{s-g} = +\infty$ days, or forever.

Chapter 7

7.1 It's easiest to let $t = 0$ correspond to the year 1800 and measure population in billions of people. Then the exponential growth law takes the form $P(t) = e^{rt}$, where r must be determined. Note that $P(0) = 1$ (billion). The condition $P(200) = 6$ implies that $r = \frac{\ln 6}{200} \doteq 0.009$ yrs^{-1}. We must find the time t at which $P(t) = e^{rt} = 10^{-9}$, where 10^{-9} is one person. Using the known value of r, we find that $t = -2313$, which means the population was 1 person 2313 years before 1800, or in 513 B.C.E.

7.3 The speeds decreased exponentially (constant percentage rate of decrease). The function $v_a(t) = 4e^{t \ln 0.9} = 4(0.9)^t$ describes Abe's speed ($v_a(0) = 4$ with a 10% decrease each hour). The function $v_s(t) = 3e^{t \ln 0.81} = 3(0.81)^t$ describes Sally's speed ($v_s(0) = 3$ with a 19% decrease each hour). Integrating, we find that Abe's distance from Felicity is given by

$$d_a(t) = \int_0^t v_a(y)\, dy = \int_0^t 4e^{y \ln 0.9}\, dy = \frac{4}{\ln 0.9}(e^{t \ln 0.9} - 1) \quad \text{for } t \geq 0.$$

Similarly, Sally's distance from Cheer is given by

$$d_s(t) = \int_0^t v_s(y)\, dy = \int_0^t 3e^{y \ln 0.81}\, dy = \frac{3}{\ln 0.81}(e^{t \ln 0.81} - 1) \quad \text{for } t \geq 0.$$

At the moment of passing, we have $d_a(t) + d_s(t) = 10$, which means

$$\frac{4}{\ln 0.9}(e^{t \ln 0.9} - 1) + \frac{3}{\ln 0.81}(e^{t \ln 0.81} - 1) = 10.$$

Note that $e^{t \ln 0.81} = (0.81)^t = (0.9)^{2t}$, which reduces the previous equation to

$$\frac{3}{\ln 0.81}u^2 + \frac{4}{\ln 0.9}u - 10 - \frac{3}{\ln 0.81} - \frac{4}{\ln 0.9} = 0,$$

where $u = (0.9)^t$. The equation has the form $au^2 + bu + c = 0$, where $a = \frac{3}{\ln 0.81} = -14.24$, $b = \frac{4}{\ln 0.9} = -37.96$, and $c = -10 - \frac{3}{\ln 0.81} - \frac{4}{\ln 0.9} = -42.20$. The roots of the quadratic equation are $u = (0.844, -3.51)$, of which only the first root is meaningful. Solving $(0.91)^t = 0.844$, we see that the time of passing was approximately $t^* = 1.606$ hours. Substituting into the distance functions, $x_a(t^*) = 5.912$ kilometers and $x_b(t^*) = 4.088$ kilometers. Abe and Sally met approximately 5.912 kilometers from Felicity.

7.5(i) Intersection of the curves $y = e^x$ and $y = x^p$ requires $e^x = x^p$. Tangency of the two curves requires that the derivatives be equal: $e^x = px^{p-1}$. A bit of algebra reveals that $p = x = e$. Thus the curves $y = e^x$ and $y = x^e$ are tangent at the point (e, e^e).

7.7(i) The amount of oil extracted between $t = 0$ and a later time t is

$$Q(t) = \int_0^t r(s)\, ds = 10^7 \int_0^t e^{-ks}\, ds = \frac{10^7}{k}(1 - e^{-kt}).$$

At this rate, the total amount of oil that could be extracted is $Q_\infty = \lim_{t \to \infty} Q(t) = \frac{10^7}{k}$. Setting Q_∞ equal to the total reserve of 2×10^9 tons, we find that $k = 0.005$ yr^{-1}.

7.7(ii) We now assume that the rate of extraction is $2r(t) = 2 \cdot 10^7 e^{-kt}$. Using the value of k found in part (i), the amount of oil extracted between $t = 0$ and a later time t is

$$Q(t) = 2 \cdot 10^7 \int_0^t e^{-0.005s}\, ds = 2 \cdot \frac{10^7}{0.005}(1 - e^{-0.005t}).$$

Solving $Q(t) = 2 \times 10^9$ tons, the total reserves will be depleted after $t = \frac{\ln 2}{0.005} = 6.0$ years.

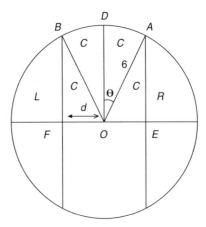

7.9(i) We need to consider only the upper half of the pizza (see figure above); the left region (L) and the right region (R) have the same area. If we can find the area of the center region (C), then the area of R and L can be found, knowing that the area of the entire upper half is $A_T = 18\pi$. Using the figure above and leaving d unspecified, $\theta = \sin^{-1}\left(\frac{d}{6}\right)$. The center region consists of the sector AOB and triangles $\triangle AOE$ and $\triangle BOF$. The area of C is

$$A_C(d) = \underbrace{\frac{1}{2} \cdot 6^2 \cdot 2\theta}_{\text{sector}} + \underbrace{2 \cdot \frac{1}{2} \cdot d\sqrt{36 - d^2}}_{\text{triangles}} = 36\theta + d\sqrt{36 - d^2}.$$

The area of the right region is $A_R(d) = \frac{1}{2}(A_T - A_C(d))$. With $d = 1.8$ inches, we find that $A_C = 21.27$ square inches and $A_R = 17.64$ square inches. The center region is larger.

7.9(ii) We now define "value functions" that consist of the area of the regions, weighted by 1, and boundary length, weighted by 1.5. The arc length of the center region is $6 \cdot 2\theta$. The arc length of the right region is $6\left(\frac{\pi}{2} - \theta\right)$. The value of the center region is $V_C(d) = A_C(d) + 1.5 \cdot 6 \cdot 2\theta$. The value of the right region is $V_R(d) = A_R(d) + 1.5 \cdot 6\left(\frac{\pi}{2} - \theta\right)$. With $d = 1.5$, we find that $V_C = 22.36$ and $V_R = 31.23$. The left and right regions (equal in value) are more valuable than the center region.

7.11(i) The volume of the can is $V = \pi r^2 h = 354$, and the surface area is $S = 2\pi r^2 + 2\pi rh$. Using the volume expression to eliminate h in the surface area expression gives $S(r) = 2\pi r^2 + \frac{2V}{r}$. The critical points of $S(r)$ satisfy $S'(r) = 0$, or $r^3 = \frac{V}{2\pi}$. With $V = 354$, the critical point is $r = 3.83$ cm., which gives $h = 7.67$ cm. ($S''(r) > 0$, so we have found a minimum.)

7.11(ii) The only change to part (i) is that we double the area of the ends of the can; thus $S = 2\pi r^2 + 4\pi rh$. Proceeding as before, we find that the optimal dimensions are $r = 3.04$ cm and $h = 12.1$ cm, which are almost exactly the dimensions of a real can.

7.13 For each angle θ, with $0 \le \theta \le \frac{\pi}{2}$, there is a pole of maximum length that just fits in the corner, touching the two exterior walls and the inside corner point P (see figure); call this length $\ell(\theta)$. Using the figure, we have

$$\ell(\theta) = x + y = \frac{3}{\sin\theta} + \frac{4}{\cos\theta}.$$

As a check, note that $\ell(0) = \ell\left(\frac{\pi}{2}\right) = \infty$, as expected. To turn the corner, a pole must fit in the position shown in the figure *for all* angles θ in the interval $0 \le \theta \le \frac{\pi}{2}$. Therefore the maximum length of the pole that fits in the corner *for all* angles θ is the minimum value of $\ell(\theta)$ as θ varies over the interval $0 \le \theta \le \frac{\pi}{2}$. To find a local minimum of $\ell(\theta)$, we differentiate:

$$\ell'(\theta) = -\frac{3\cos\theta}{\sin^2\theta} + \frac{4\sin\theta}{\cos^2\theta} = 0.$$

Note that $\theta = 0$ and $\theta = \frac{\pi}{2}$ cannot lead to a local minimum, so we can assume that $\sin\theta \neq 0$ and $\cos\theta \neq 0$. The critical point satisfies $3\cos^3\theta = 4\sin^3\theta$ or $\tan^3\theta = \frac{3}{4}$. Solving, we find that $\theta^* = \tan^{-1}\left(\sqrt[3]{3/4}\right) = 0.738$ radians. (θ^* corresponds to a local minimum.) The pole of maximum length that turns the corner has length $\ell(\theta^*) = 9.87$ feet.

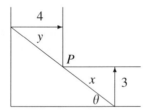

7.15 The volume of water needed to cover a marble of radius r is the volume of the cylinder of water minus the volume of the marble, or $V(r) = \pi \cdot 4^2 \cdot (2r) - \frac{4}{3}\pi r^3$, where we have used the fact that the depth of the water is twice the radius of the marble. Thus $V(r) = 32\pi r - \frac{4}{3}\pi r^3$. To maximize the volume function, we differentiate and find the critical points.

It follows that $V'(r) = 32\pi - 4\pi r^2 = 0$, or $r^* = 2\sqrt{2} \doteq 2.83$ inches. Note that $V''(r^*) < 0$, so we have indeed found a local maximum. A marble of radius 2.82 inches requires the most water to cover it.

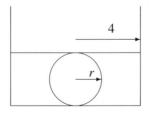

7.17 Let P be the point on shore nearest the boat as the woman begins rowing. Suppose that she lands the boat at a point x miles from P and $6 - x$ miles from the restaurant. Then she rows $\sqrt{x^2 + 16}$ miles and walks $6 - x$ miles. The travel time for the entire trip is $t(x) = \frac{\sqrt{x^2+16}}{v} +$

$\frac{6-x}{3}$, where v is her rowing speed. To minimize the travel time, we solve $t'(x) = 0$ to find the critical points. The critical points satisfy $\frac{3x}{\sqrt{x^2+16}} = v$. We must find the value of v such that $x = 6$ (no walking) is a critical point. Substituting $x = 6$, we find that $v = \frac{9}{\sqrt{13}}$ miles per hour.

7.19(i) All three parts can be solved with the same formulation. Assume that the original piece of cardboard has side lengths a and $b \geq a$. Squares with sides of length $x \leq a$ are cut from the corners of the rectangle. The resulting box has a volume $V(x) = x(a - 2x)(b - 2x) = 4x^3 - 2(a+b)x^2 + abx$. To find the critical points, we solve $V'(x) = 12x^2 - 4(a+b)x + ab = 0$, which has roots $x = \frac{1}{6}((a + b) \pm \sqrt{a^2 - ab + b^2})$. With $a = 3$ and $b = 4$, the critical points are $x = \frac{1}{6}(7 \pm \sqrt{13})$, of which only the smaller root is meaningful; it also corresponds to a maximum of the volume function. The resulting box has a volume of $V\left(\frac{1}{6}(7 - \sqrt{13})\right) \doteq 3.03$ cubic feet.

7.19(ii) Using the results of part (i) with $a = b$, the critical points are $x = \frac{a}{2}, \frac{a}{6}$, of which the smaller root corresponds to a maximum volume. The resulting box has a volume of $\frac{2a^3}{27}$ cubic feet.

7.19(iii) From part (i), the relevant critical point that maximizes the volume is $x(a, b) = \frac{1}{6}((a + b) - \sqrt{a^2 - ab + b^2})$. Holding a fixed, we find that

$$\lim_{b \to \infty} x(a, b) = \lim_{b \to \infty} \left[\frac{a}{6} + \frac{b}{6} \left(1 - \sqrt{\left(\frac{a}{b}\right)^2 - \left(\frac{a}{b}\right) + 1} \right) \right] = \frac{a}{4}.$$

(Expanding the square root using the binomial theorem is easier than using L'Hôpital's rule.) In the limit of long thin rectangles, the corner squares should have side lengths of $\frac{a}{4}$.

7.21 Suppose that the ant lands a distance r from the center of the circle and begins crawling at an angle θ measured counterclockwise from the positive horizontal direction. By symmetry, the average length of its path, $A(r)$, depends only on r. Let $d(r, \theta)$ be the distance from the landing point to the edge of the circle in the direction θ. Note that $d(r, \theta) + d(r, \theta + \pi)$ is the length of the chord, which is $2\sqrt{R^2 - r^2} \sin \theta$. The average length of the path over all angles $0 \leq \theta \leq 2\pi$ is

$$A(r) = \frac{1}{2\pi} \int_0^{2\pi} d(r, \theta) \, d\theta = \frac{1}{2\pi} \int_0^{\pi} d(r, \theta) + d(r, \theta + \pi) \, d\theta$$

$$= \frac{1}{\pi} \int_0^{\pi} \sqrt{R^2 - r^2} \sin \theta \, d\theta$$

$$= \frac{2}{\pi} \int_0^{\pi/2} \sqrt{R^2 - r^2} \sin \theta \, d\theta.$$

We have found the average path length for one landing point within the circle. To find the average path length over all landing points, we must find the average value of $A(r)$ over the interior of the circle. The average path length is

$$\overline{A} = \frac{1}{\pi r^2} \int_0^{2\pi} \int_0^R A(r) \, r \, dr \, d\alpha$$

$$= \frac{1}{\pi r^2} \int_0^{2\pi} \int_0^R \left(\frac{2}{\pi} \int_0^{\pi/2} \sqrt{R^2 - r^2} \sin \theta \, d\theta \right) r \, dr \, d\alpha.$$

Some computation leads to the intermediate result

$$\overline{A} = \frac{4R}{3\pi} \int_0^{\pi/2} \frac{1 - \cos^3 \theta}{\sin^2 \theta} \, d\theta.$$

Evaluating this improper integral carefully at $\theta = 0$, we find the average path length $A = \frac{8R}{3\pi}$.

Chapter 8

8.1 Let P_n be the population at n years, with $P_0 = 650$. The governing difference equation is $P_{n+1} = (1 + 0.05 - 0.035)P_n = 1.015P_n$. The solution is $P_n = 650(1.015)^n$ for $n = 0, 1, 2, \ldots$. The population first exceeds 2500 when $n = 91$ years.

8.3 Let B_n be the balance in the account at n months, with $B_0 = 0$. The annual interest rate of 9.0% corresponds to a monthly rate of $(9.0/12)\% = 0.0075$. The governing difference equation is $B_{n+1} = 1.0075B_n + 100$. As shown in Example 8.2, the solution is $B_n = 100\frac{(1.0075)^n - 1}{0.0075}$ for $n = 0, 1, 2, \ldots$. Solving $B_n = \$20,000$ for n, we find that it takes $n = 123$ months to reach a balance of \$20,000.

8.5 Let B_n be the balance in the loan at n months, with $B_0 = \$150,000$. Let p be the monthly payment. The annual interest rate of 6.0% corresponds to a monthly rate of $(6.0/12)\% = 0.005$. The governing difference equation is $B_{n+1} = 1.005B_n - p$. As shown in Example 8.2, the solution is

$$B_n = 150,000(1.005^n) - p\frac{1.005^n - 1}{0.005}, \quad \text{for } n = 0, 1, 2, \ldots.$$

Setting $B_n = 0$ and $n = 30 \cdot 12 = 360$ months, the required monthly payment is $p = \$899$.

8.7 Let $n = 0$ correspond to *today*. The weight of the pig on the nth day after today is $w_n = 200 + 5n$. The price per pound on the nth day is $p_n = 65 - n$. The amount spent on food by the nth day is $f_n = 45n$. Therefore, the market value of the pig on the nth day is $v_n = w_n p_n - f_n = (200 + 5n)(65 - n) - 45n = -5n^5 + 80n + 13,000$. Enumerating or plotting this sequence, it reaches a maximum value at $n = 8$ days, when the value of the pig is about \$133.

8.9 Letting x_n be the number of hours of sleep on the nth night, the assumption of the model gives the difference equation $x_{n+1} = \frac{1}{2}(x_n + x_{n-1})$ for $n = 1, 2, 3, \ldots$. The initial conditions are $x_0 = 7$ and $x_1 = 8$. Using a trial solution of the form $x_n = a^n$ leads to the characteristic polynomial $a^2 - \frac{1}{2}a - \frac{1}{2} = 0$, which has roots $a = 1$ and $a = -\frac{1}{2}$. Thus, the general solution of the difference equation is $x_n = c_1 \cdot 1^n + c_2 \left(-\frac{1}{2}\right)^n$. The initial conditions imply that $c_1 = \frac{23}{3}$ and $c_2 = -\frac{2}{3}$. The solution to the initial value problem is

$$x_n = \frac{23}{3} - \frac{2}{3}\left(-\frac{1}{2}\right)^n.$$

Letting $n \to \infty$, we see that $x_n \to \frac{23}{3}$. Thus, in the long run, if the pattern persists, you will get close to $7\frac{2}{3}$ hours of sleep per night.

8.11 The difference equation for the Fibonacci sequence is $F_{n+1} - F_n - F_{n-1} = 0$. Now divide this equation through by F_n and let $r_n = F_n/F_{n-1}$. The result is $r_{n+1} - 1 - (1/r_n) = 0$. Assuming that $\phi = \lim_{n\to\infty} r_n$ exists, take the limit across this equation. The result is the polynomial equation $\phi^2 - \phi - 1 = 0$. The relevant (positive) root is $\phi = \frac{1+\sqrt{5}}{2}$, the golden mean.

8.13 Consider a group of N families. With the one-child policy, these N families will have one child each, for a total of N children. Under the one-son policy, $\frac{N}{2}$ families will have a boy, which will be their last child. However, $\frac{N}{2}$ families will have a girl and can have another child. Of these $\frac{N}{2}$ families, $\frac{N}{4}$ families will have a boy, which will be their last child. However, $\frac{N}{4}$ families will have a girl and can have another child. If this pattern continues, there will be $\frac{N}{2} + \frac{N}{4} + \frac{N}{8} + \ldots$ boys and the same number of girls. Recalling the geometric series, we see that $\frac{N}{2} \sum_{k=0}^{\infty} \left(\frac{1}{2}\right)^k = \frac{N}{2} \cdot 2 = N$. Thus in the long run there will be N boys and N girls, or twice as many children as under the one-child policy. *More directly*, note that eventually all N families will have a boy. Assuming that the natural birth rate prevails (half boys and half girls), there will be as many girls as boys in the long run. Thus there will be N boys and N girls.

8.15 Let $h = 10$ meters be the initial height. The time to fall to the ground the first time is $\sqrt{\frac{2h}{g}}$. The ball rebounds to a height of ph, and the time to rise to that height and fall from that height is $\sqrt{\frac{2ph}{g}}$ (each way). The ball next rebounds to a height of $p^2 h$, and the time to rise to that height and fall from that height is $\sqrt{\frac{2p^2 h}{g}}$ (each way). Continuing in this way for each bounce, the total time for the ball to come to rest is

$$T = \sqrt{\frac{2h}{g}} + 2 \sum_{k=1}^{\infty} \sqrt{\frac{2p^k h}{g}} = \sqrt{\frac{2h}{g}} \left(1 + 2 \sum_{k=1}^{\infty} (\sqrt{p})^k \right).$$

Evaluating the geometric series and letting $h = 10$ and $g = 9.8$, we have $T = 1.43 \left(\frac{1+\sqrt{p}}{1-\sqrt{p}}\right)$ seconds. Note that $T \to \infty$ as $p \to 1$, and $T \to 1.43$ as $p \to 0$ (which is the fall time with no bounce).

8.17(i) Let the fractions of the population in the highlands and lowlands in month n be H_n and L_n, respectively. The transition rates lead to the system of difference equations

$$\begin{pmatrix} H_{n+1} \\ L_{n+1} \end{pmatrix} = \begin{pmatrix} 0.6 & 0.2 \\ 0.4 & 0.8 \end{pmatrix} \begin{pmatrix} H_n \\ L_n \end{pmatrix} \quad \text{for } n = 0, 1, 2, 3, \ldots,$$

where $H_0 = 0.9$ and $L_0 = 0.1$.

We look for solutions of the form

$$\begin{pmatrix} H_n \\ L_n \end{pmatrix} = \mathbf{v}\lambda^n = \begin{pmatrix} v_1 \\ v_2 \end{pmatrix} \lambda^n,$$

where λ and $\mathbf{v}$ must be determined. Substitution of the trial solution into the difference equation results in the characteristic polynomial for λ:

$$\det \begin{pmatrix} 0.6 - \lambda & 0.2 \\ 0.4 & 0.8 - \lambda \end{pmatrix} = \lambda^2 - 1.4\lambda + 0.4 = 0.$$

The roots of the polynomial are the eigenvalues $\lambda_1 = 1$ and $\lambda_2 = 0.4$. For each eigenvalue, we must find the associated eigenvector. For $\lambda_1 = 1$, the eigenvectors are $c_1(1, 2)^T$, and for $\lambda_2 = 0.4$, the eigenvectors are $c_2(1, -1)^T$, where c_1 and c_2 are arbitrary constants. Thus the general solution is

$$\begin{pmatrix} H_n \\ L_n \end{pmatrix} = c_1 \begin{pmatrix} 1 \\ 2 \end{pmatrix} 1^n + c_2 \begin{pmatrix} 1 \\ -1 \end{pmatrix} 0.4^n.$$

The initial conditions imply that $c_1 = \frac{1}{3}$ and $c_2 = \frac{17}{30}$. The solution to the initial value problem is

$$\begin{pmatrix} H_n \\ L_n \end{pmatrix} = \frac{1}{3} \begin{pmatrix} 1 \\ 2 \end{pmatrix} + \frac{17}{30} \begin{pmatrix} 1 \\ -1 \end{pmatrix} 0.4^n \quad \text{for } n = 0, 1, 2, \ldots.$$

8.17(ii) Using the solution of part (i), as $n \to \infty$, the deer populations reach a steady state distribution of $\frac{1}{3} : \frac{2}{3}$; that is, there will be twice as many deer in the lowlands as in the highlands.

Chapter 9

9.1 The ordering is $A \geq G \geq H$, with equality only if the n numbers are equal.

9.7 (i) With an interest rate of 0.5%, the balances are $69,306 and $298,724. (ii) With an interest rate of 0.7%, the balances are $92,877 and $588,304. With an interest rate of 0.3%, the balances are $52,611 and $160,580. (iii) All balances are 10% greater if the monthly deposit is 10% greater.

9.9 The open lockers are perfect squares $(1, 4, 9, 16, \ldots)$.

9.11 Note that the iteration $x_{n+1} = g(x_n) = x_n{}^a$ describes building the tower from the top, and the iteration $x_{n+1} = h(x_n) = a^{x_n}$ describes building the tower from the bottom, where we take $x_0 = a$. If the sequences generated by g and h converge, they converge to a fixed point, that is, a point that satisfies $x = g(x)$ and $x = h(x)$. For $a > 1$, the only fixed point of g is $x = 1$, and the sequences generated diverge for $a > 1$. To find the fixed points of h, we first find the value of a such that the curves $y = x$ and $y = a^x$ are tangent. Intersection of the curves $y = a^x$ and $y = x$ requires that $a^x = x$. Tangency of the two curves requires that the derivatives be equal, or $(\ln a)a^x = 1$. These two conditions imply that $x = \frac{1}{\ln a}$. A bit of algebra leads to $\ln a = 1/e$, or $a = e^{1/e} \doteq 1.445$. It then follows that $x = e$. Thus h has fixed points, provided that $1 < a < e^{1/e}$, and the tower has a finite value. The actual value of the fixed point is the smaller root of $a^x = x$, which must be approximated numerically, say, by Newton's method. The values of the tower for $1 < a < e^{1/e}$ are shown in the figure below.

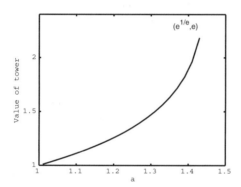

9.13 The man walks for approximately 2.72 hours or 3.26 miles.

9.15 Referring to the figure on the next page, let $x = a + b$ be the unknown width of the alley. Using the right triangle with hypotenuse of length 20, we have a pair of similar triangles with

$$\frac{a}{c} = \frac{x}{20}.$$

Using the right triangle with hypotenuse of length 25, we have a pair of similar triangles with

$$\frac{b}{d} = \frac{x}{25}.$$

Using the Pythagorean theorem on the two lower, smaller, right triangles, we also have

$$a^2 + 36 = c^2 \quad \text{and} \quad b^2 + 36 = d^2.$$

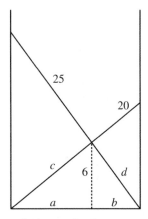

Reducing these equations leaves a single equation for x:

$$100 = \frac{5 \cdot 120}{\sqrt{400 - r^2}} + \frac{4 \cdot 150}{\sqrt{625 - r^2}}$$

or

$$f(x) = \frac{1}{\sqrt{400 - x^2}} + \frac{1}{\sqrt{625 - x^2}} - \frac{1}{6} = 0.$$

This equation cannot be solved analytically. Proceeding with Newton's method or a graphical solution, the width of the alley is approximately $x = 17.8$ feet.

9.17 It suffices to consider a two-dimensional cross-section of the sphere (a circle of radius r) and the cone (a triangle of height 1). Let $\theta = \frac{\pi}{12}$ be half of the cone angle. Let a be the distance between the center of the circle and the top of the cone, with $a > 0$ when the center is above the cone and $a < 0$ when the center is below the top of the cone. From the figure below, we see that in either case, $r = (1 + a)\sin\theta$ or $a = \frac{r}{\sin\theta} - 1$. Note that $a_{max} = \tan^2\theta$, which implies that $r_{max} = \sec^2\theta\sin\theta \doteq 0.277$. At the other endpoint, $r_{min} = -a_{min} = \frac{1}{1+\csc\theta} \doteq 0.205$.

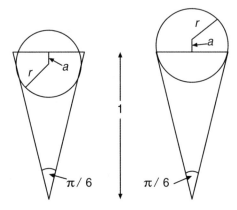

Recall that the area of a cap of a circle of radius r subtended by a central angle ϕ is $\frac{1}{2}r^2(\phi - \sin\phi)$. When the center of the circle is above the top of the cone ($a > 0$), the area of the region below the top of the cone is $A_1 = \frac{1}{2}r^2(\phi - \sin\phi)$, where $\phi = 2\cos^{-1}\left(\frac{a}{r}\right)$. When the center of the circle is below the top of the cone ($a < 0$), the area of the region below the top of the cone is $A_2 = \pi r^2 - \frac{1}{2}r^2(\phi - \sin\phi)$, where $\phi = 2\cos^{-1}\left(\frac{|a|}{r}\right)$. The figure on the next page shows the graph of the area of the region below the top of the cone as a function of r, where

$r_{min} \leq r \leq r_{max}$. We see that the maximum area occurs when the radius is approximately $r = 0.216$ units.

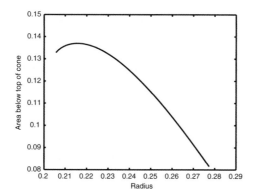

9.19 Let $t = 0$ be the time at which the snow started falling, and let $t = T$ be 12:00. The snow depth is $d(t) = kt$, where k is a constant. The speed of the plow is $s(t) = a/(d(t))^3 = a/(kt)^3$, where a is a constant. Integrating and imposing the conditions of the problem, we have

$$\int_T^{T+1} \frac{a}{(kt)^3} \, dt = 3 \int_{T+1}^{T+2} \frac{a}{(kt)^3} \, dt.$$

Simplifying the integrals leads to the cubic equation $T^3 - 3T - 1 = 0$. Numerical or graphical approximation shows that the relevant root is $T \doteq 1.88$ hours. Thus the snow started falling approximately 1.88 hours before 12:00.

9.21 The speed function that decreases by 1% every mile is $v_A(x) = \frac{dx}{dt} = v_0 e^{ax}$, where $a = \ln 0.99 \text{ mile}^{-1}$ (note $a < 0$). The speed function that decreases by 1% every minute is $v_B(t) = \frac{dx}{dt} = v_0 e^{at}$, where $a = \ln 0.99 \text{ minute}^{-1}$. Let L be the length of the trip. In case A, we can integrate the velocity function and solve for the travel time:

$$\int_0^L e^{-ax} \, dx = \int_0^{t_A} v_0 \, dt \quad \text{or} \quad t_A(L) = \frac{1}{av_0}(1 - (0.99)^{-L}).$$

In case B, we can integrate and solve to get

$$\int_0^L dx = v_0 \int_0^{t_B} e^{at} \, dt \quad \text{or} \quad t_B(L) = \frac{1}{a} \ln \left(1 + \frac{aL}{v_0}\right).$$

Now the required comparisons can be made. (i) With $v_0 = 0.5$ and $L = 40$, we find that $t_A = 98.5$ minutes and $t_B = 162.1$ minutes, so (A) is faster. Note that $t_B \to \infty$ as $L \to -\frac{v_0}{a} \doteq 50$ miles. (ii) With $v_0 = 2$ and $L = 100$, we find that $t_A = 86.1$ minutes and $t_B = 69.5$ minutes, so (B) is faster. Note that $t_B \to \infty$ as $L \to -\frac{v_0}{a} \doteq 200$ miles. (iii) With $v_0 = 2$ and $L = 180$, we find that $t_A = 254.0$ minutes and $t_B = 233.7$ minutes, so (B) is faster. (iv) These results suggest that the faster strategy depends on L and v_0 in a subtle way. By graphing t_A and t_B for various values of L and v_0, you can confirm that $t_A < t_B$ for all L when $v_0 < 1$. However, for $v_0 > 1$, there is a critical value L^* (that depends on v_0) such that $t_A > t_B$ for $L < L^*$ and $t_A < t_B$ for $L > L^*$. The critical value $v_0 = 1$ can also be determined by expanding t_A and t_B in a Taylor series near $L = 0$. The determination of the regions in the (v_0, L) parameter plane for which $t_A > t_B$ and $t_A < t_B$ is a challenging numerical exercise.

Chapter 10

10.1(i) The four students from Ohio can be selected in $\binom{7}{4} = 35$ ways. Four students can be selected from the entire group in $\binom{16}{4} = 1820$ ways. Therefore the probability that all four students are from Ohio is $\frac{35}{1820} = \frac{1}{52}$.

10.1(ii) The two students from Colorado can be selected in $\binom{4}{2} = 6$ ways, and the two students from Utah can be selected in $\binom{5}{2} = 10$ ways. The four students can be selected in $\binom{16}{4} = 1820$ ways. Therefore the probability that two students are from Colorado and two students are from Utah is $\frac{6 \cdot 10}{1820} = \frac{3}{91}$.

10.1(iii) The complement of at least one of the four students being from Ohio is that none of the four students is from Ohio. The probability that none of the four students is from Ohio is $\frac{9}{16} \cdot \frac{8}{15} \cdot \frac{7}{14} \cdot \frac{6}{13} = \frac{9}{130}$. Thus the probability that at least one of the four students is from Ohio is $1 - \frac{9}{130} = \frac{121}{130}$.

10.1(iv) This event could occur if all students are from Colorado, or all are from Utah, or all are from Ohio. The number of ways that all students could be from Colorado is $\binom{4}{4} = 1$, the number of ways that all students could be from Utah is $\binom{5}{4} = 5$, and the number of ways that all students could be from Ohio is $\binom{7}{4} = 35$. The four students can be selected in $\binom{16}{4} = 1820$ ways. Thus the probability that all four students are from the same state is $\frac{1+5+35}{1820} = \frac{41}{1820}$.

10.1(v) To select at least one student from each state, it is necessary to select two from one state and one from each of the other states. There are $\binom{4}{2} \cdot 5 \cdot 7 = 210$ ways to select two students from Colorado and one from Utah and Ohio. There are $\binom{5}{2} \cdot 4 \cdot 7 = 280$ ways to select two students from Utah and one from Colorado and Ohio. There are $\binom{7}{2} \cdot 4 \cdot 5 = 420$ ways to select two students from Ohio and one from Colorado and Utah. Therefore the probability of selecting at least one student from each state is $\frac{210+280+420}{1820} = \frac{1}{2}$.

10.3(i) For a given card value (say Jack), there are $\binom{4}{3}$ ways to form three of a kind and $\binom{4}{2}$ ways to form a pair (because there are four suits). For the pair there are 13 different card values, which leaves 12 different card values for the three of a kind. Thus the probability of a full house is

$$\frac{13\binom{4}{3} \cdot 12\binom{4}{2}}{\binom{52}{5}} \doteq 0.00144.$$

10.3(ii) There are 13 different card values and hence 13 ways to have four of a kind. There are 48 ways for the fifth card to be dealt. Thus the probability of four of a kind is

$$\frac{13 \cdot 48}{\binom{52}{5}} \doteq 0.000240.$$

10.3(iii) For each card value, there are $\binom{4}{3}$ ways to form three of a kind, and there are 13 card values. The remaining two cards must have a value different from each other and from the value used for the three of a kind. So there are $\binom{12}{2}$ ways to select these two cards so they have different values and 4^2 different arrangements of suits for these two cards. Thus the probability of three of a kind and no better is

$$\frac{13\binom{4}{3} \cdot 4^2\binom{12}{2}}{\binom{52}{5}} \doteq 0.0211.$$

10.3(iv) For each card value, there are $\binom{4}{2}$ ways to form a pair, and there are 13 card values. The remaining three cards must have a value different from each other and from the value used for the pair. So there are $\binom{12}{3}$ ways to select these three cards so that they have different values, and 4^3 different arrangements of suits for these three cards. Thus the probability of a pair and no better is

$$\frac{13\binom{4}{2} \cdot 4^3 \binom{12}{3}}{\binom{52}{5}} \doteq 0.423.$$

10.3(v) In terms of card values, there are 10 different straights (5-high up to ace-high). There are 4^5 different arrangements of suits for these 10 straights. There are 40 straight flushes that must be excluded. So the probability of a straight and no better is

$$\frac{10 \cdot 4^5 - 40}{\binom{52}{5}} \doteq 0.00392.$$

10.3(vi) For each of four suits, there are $\binom{13}{5}$ ways to arrange the 13 cards of the suit. Excluding the 40 straight flushes, the probability of a flush and no better is

$$\frac{4\binom{13}{5} - 40}{\binom{52}{5}} \doteq 0.00197.$$

10.5 The probability of rolling at least a 5 is $\frac{1}{3}$. The probability of not rolling at least a 5 is $\frac{2}{3}$. The probability of rolling at least a 5 exactly zero times in 10 throws is $\left(\frac{2}{3}\right)^{10}$. The probability of rolling at least a 5 exactly one time in 10 throws is $\binom{10}{1}\left(\frac{2}{3}\right)^9 \cdot \frac{1}{3}$. Thus the probability of rolling at least a five at least twice (the complement event) is $1 - \left(\frac{2}{3}\right)^{10} - \binom{10}{1}\left(\frac{2}{3}\right)^9 \cdot \frac{1}{3} \doteq 0.896$.

10.7 It's easiest to compute the probability of the complement of the event in question: that the top three cards do not contain a royalty card. There are 12 royalty cards, $\binom{52-12}{3} = \binom{40}{3}$ ways to select three nonroyalty cards, and $\binom{52}{3}$ ways to select three cards from the full deck. So the probability that the top three cards are nonroyalty cards is $\binom{40}{3}/\binom{52}{3} \doteq 0.447$. Therefore the probability that at least one of the top three cards is a royalty card is $1 - 0.477 \doteq 0.553$.

10.9 The probability of rolling a 7 with two dice is $\frac{1}{6}$. Letting W be a win and L be a loss, the person who rolls first can win with a sequence W (with probability $\frac{1}{6}$), LLW (with probability $\left(\frac{5}{6}\right)^2 \cdot \frac{1}{6}$), LLLLW (with probability $\left(\frac{5}{6}\right)^4 \cdot \frac{1}{6}$), and so forth. Thus the probability that the first person to roll wins is

$$\frac{1}{6}\sum_{k=0}^{\infty}\left(\frac{5}{6}\right)^{2k} = \frac{1}{6}\left(\frac{1}{1-\left(\frac{5}{6}\right)^2}\right) = \frac{6}{11},$$

where the geometric series with ratio $\left(\frac{5}{6}\right)^2$ has been evaluated.

10.11 Let p be the probability of a false test on a single mammogram. Then the probability of at least one false test in ten mammograms is $1 - (1-p)^{10} = 0.5$. Solving for p, we find that $p \doteq 0.0670$.

10.13 An odd person out will be determined if $N-1$ people toss a head and one person tosses a tail, or vice versa, each of which has the same probability. The probability of $N-1$ heads (probability $\left(\frac{1}{2}\right)^{N-1}$) and one tail (probability $\frac{1}{2}$) is

$$N \cdot \left(\frac{1}{2}\right)^{N-1} \cdot \frac{1}{2} = N \cdot \left(\frac{1}{2}\right)^{N}.$$

Therefore the probability of an odd person out (with either a head or a tail) on a single trial is twice the above probability, or $p = N \left(\frac{1}{2}\right)^{N-1}$. The probability that it takes exactly k trials to determine an odd person out is $P_k = p(1-p)^{k-1}$ (using Bernoulli trials). The average number of tosses needed to determine an odd person out is the weighted average

$$A(N) = \sum_{k=1}^{\infty} kP_k = p \underbrace{\sum_{k=1}^{\infty} k(1-p)^{k-1}}_{S}.$$

The last series, $S = 1 + 2(1-p) + 3(1-p)^2 + \cdots$, can be valuated by multiplying both sides by $(1-p)$ and subtracting the two expressions. The result is a geometric series that can be simplified to yield $S = 1/p^2$. Combining these results, we find that $A(N) = pS = \frac{1}{p} = 2^{N-1}/N$. With N people in the group, the expected (average) number of coin tosses needed to determine an odd person out is $2^{N-1}/N$.

10.15 (i) floor(42*rand+1), (ii) 3 + 3*rand.

10.17 The region in the unit square containing points (x, y) such that $|x - y| < \frac{1}{2}$ is the region in the unit square between the lines $y = x - \frac{1}{2}$ and $y = x + \frac{1}{2}$. By integration or geometry, the area of this region is $\frac{3}{4}$. The area of the unit square is 1. Thus the required probability is $\frac{3/4}{1} = \frac{3}{4}$.

10.19 A bit of geometry shows that the side length of the equilateral triangle is $\sqrt{3}$. Random chord (A) is generated as in Example 10.7, where it was found that the probability that a random chord has length *less* than r is $p(r) = \frac{2}{\pi} \sin^{-1}\left(\frac{r}{2}\right)$. Therefore the probability that a random chord generated by method (A) has a length greater than $\sqrt{3}$ is $1 - p(\sqrt{3}) = \frac{1}{3}$.

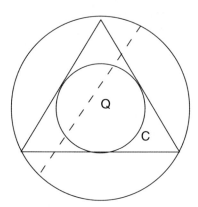

For method (B), a picture is useful. As shown in the figure, a random chord has a length greater than the side length of the triangle, provided that the center of the chord lies inside the inscribed circle C. This circle has a radius of $\frac{1}{2}$, so its area is $\frac{\pi}{4}$. The area of the outer (unit) circle is π. Therefore the probability that a random chord generated by method (B) has a length greater than $\sqrt{3}$ is $\frac{\pi/4}{\pi} = \frac{1}{4}$.

10.21 The dart board is shown in the figure on the next page. Without loss of generality, we can assume that the circle has unit radius. The probability that a randomly thrown dart lands inside the circle is the ratio of the area of the circle to the area of the triangle. The area of the unit circle is π. So the task reduces to finding the area of the triangle, which amounts to finding the length of its sides, x. Referring to the figure, we can use the Pythagorean theorem twice:

$$\left(\frac{x}{2}\right)^2 + 1 = y^2 \quad \text{and} \quad \left(\frac{x}{2}\right)^2 + (1+y)^2 = x^2.$$

Eliminating y from the first of these equations and solving the second equation for x gives us $x = 2\sqrt{3}$. Thus, the height of the triangle is $1 + y = 3$, and the area of the triangle is $3\sqrt{3}$. This means the probability that a randomly thrown dart lands inside the circle is $\frac{\pi}{3\sqrt{3}} \doteq 0.605$.

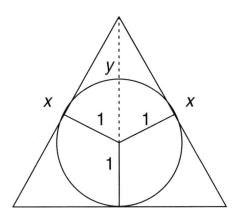

10.23 Label the sides of the triangle and their lengths a, b, and c. If the triangle is equilateral ($a = b = c$), then each side length is the mean of the other side lengths. If the triangle is isosceles with $a = b < c$, then a is less than the mean of b and c. Now with a fixed, imagine deforming an isosceles triangle into any triangle with $a < b < c$. Clearly, the mean of b and c increases, while a remains fixed. Therefore, a is less than the mean of b and c. In all triangles, the length of one side is always less than or equal to the mean of the other two side lengths, so the probability in question is 1.

10.25 Assume that the square is in the first quadrant with one vertex at $(0, 0)$. Without loss of generality, assume that the first point falls on the lower side of the square. Then there are four possibilities, each with probability $\frac{1}{4}$: (1) the second point falls on the lower side of the square (in which case the resulting chord has length less than 1), (2) the second point falls on the left side of the square (in which case the resulting chord may have length less than 1), (3) the second point falls on the right side of the square (in which case the resulting chord may have length less than 1), and (4) the second point falls on the upper side of the square (in which case the resulting chord cannot have length less than 1). We must consider cases 2 and 3, which are identical by symmetry.

Let $(x, 0)$ be the coordinates of the first point on the lower side of the square, and let $(0, y)$ be the coordinates of the second point on the left side of the square. The chord between these points has length $\sqrt{x^2 + y^2}$. Thus, the chord has length less than 1 (a hit), provided that $x^2 + y^2 < 1$ or that $y < \sqrt{1 - x^2}$. As shown in the figure on the next page, the region corresponding to a hit lies below the arc of the unit circle and has an area of $\frac{\pi}{4}$. The same analysis applies when the second point lies on the right side of the square. Thus the probability that the resulting chord has a length less than 1 consists of four contributions, corresponding to the side on which the second point lies:

$$p = \underbrace{\frac{1}{4}}_{\text{lower side}} + \underbrace{\frac{1}{4} \cdot \frac{\pi}{4}}_{\text{left side}} + \underbrace{\frac{1}{4} \cdot \frac{\pi}{4}}_{\text{right side}} + \underbrace{0}_{\text{upper side}} = \frac{1}{4} + \frac{\pi}{8} \doteq 0.643.$$

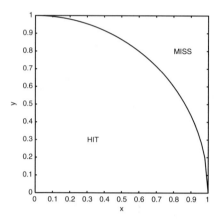

10.27 The question could also be phrased, what is the probability that three random points inside a circle lie on the same side of a diameter of the circle? Without loss of generality, we can assume that the first point P_1 falls on the positive x-axis (see figure below). Assume that the second point P_2 falls with a central angle θ with respect to the first point; notice that only the *angle* between the points is relevant. As shown in the figure, to place all three points on the same side of a diameter the third point can land anywhere inside of the sector bounded by the diameter passing through P_1 and the diameter passing through P_2. Thus, if P_3 lands anywhere inside a sector subtended by the angle $2\pi - \theta$ (marked *Hit* in the figure), all three points lie inside of a semicircle. For a given angle θ, the probability of a hit is $p(\theta) = \frac{2\pi-\theta}{2\pi}$. To find the probability of a hit over all angles $0 \leq \theta \leq \pi$, we integrate:

$$P = \frac{1}{\pi} \int_0^\pi \frac{2\pi - \theta}{2\pi} \, d\theta = \frac{3}{4}.$$

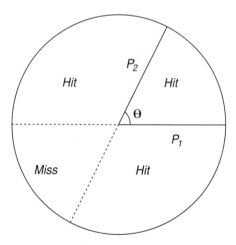

10.29 Identify the stick with the unit interval, and let the two break points be $0 < x < y < 1$. The lengths of the resulting sticks are $a = x$, $b = y - x$, and $c = 1 - y$. In order to form a triangle from the three sticks, we must have $a < b+c$, $b < a+c$, and $c < a+b$. Written in terms of x and y, the conditions become $y > \frac{1}{2}$, $y < x + \frac{1}{2}$, and $x < \frac{1}{2}$. This region has an area of $\frac{1}{8}$. Because we assumed that $x < y$, the area of the sample space is $\frac{1}{2}$. Thus the probability of forming a triangle is $\frac{1/8}{1/2} = \frac{1}{4}$.

10.31(i) Let the positions of the n points be $0 < x_i < 1$, for $i = 1, \ldots, n$. The volume of the sample space is 1. The conditions of the problem are $x_1 > 0.6$ and $x_i < x_1$, for $i = 2, \ldots, n$. The volume of this region (and the probability we seek) is given by

$$\int_{0.6}^{1} dx_1 \int_{0}^{x_1} dx_2 \cdots \int_{0}^{x_1} dx_n = \int_{0.6}^{1} x_1^{n-1} \, dx_1 = \frac{1}{n}(1 - 0.6^n).$$

The probability of meeting the conditions tends to zero as the number of points increases.

Chapter 11

11.1(i) Let $c = 0.75\pi$ yards be the circumference of the wheels, L be the distance traveled, and p be the percentage increase in energy used per gear. Let $r_k = 1 + 0.1k$ be the gears, for $k = 0, \ldots, 20$. The number of pedal strokes required to ride L yards in gear r_k is $N_k = \frac{L}{cr_k}$. The energy required per stroke in gear k is $e_k = (1 + p)^k$. Thus the energy required to ride L yards in gear k is

$$E_k = N_k e_k = \frac{L}{cr_k}(1 + p)^k.$$

Plotting this function with the given parameter values, we see that the total energy is minimized with $r_k = 1.7$ (eighth gear). The total energy used is about 375 units.

11.1(ii) Using the formulation of part (i), we see that the distance traveled in gear k, given a fixed amount of energy E, is

$$L_k = \frac{Ecr_k}{(1 + p)^k}.$$

Plotting this function with the given parameter values, the total distance is maximized with $r_k = 1.7$ (eighth gear). The total distance traveled is about 4000 yards.

11.3 Imagine the soda confined within a weightless can, which allows us to neglect the can. The main fact that we need is the old teeter-totter principle: If a person of mass m_1 is located at a coordinate x_1 and another person of mass m_2 is located at a coordinate x_2, the balance point (center of gravity) has the coordinate $\overline{x}$, where $m_1(x_1 - \overline{x}) = m_2(x_2 - \overline{x})$. Solving this equation for the position of the center of gravity, we have

$$\overline{x} = \frac{m_1 x_1 + m_2 x_2}{m_1 + m_2}.$$

Assume that the bottom of the can is at $x = 0$ and that the soda level is $x = h$, where $0 \le h \le L$. We let r be the radius of the can, ρ_ℓ be the density of the soda, and ρ_a be the density of air. The mass of the soda is $\pi r^2 h \rho_\ell$, and the mass of the air is $\pi r^2 (L - h) \rho_a$. The cylinder of soda can be localized at its midpoint, which is $x = h/2$ (that is, the soda acts like a point mass at $x = h/2$). Likewise, the cylinder of air can be localized at its midpoint, which is $x = (L + h)/2$. Thus the center of gravity is

$$\overline{x} = \frac{\pi r^2 h \rho_\ell \cdot \frac{h}{2} + \pi r^2 (L - h) \rho_a \cdot \frac{(L+h)}{2}}{\pi r^2 h \rho_\ell + \pi r^2 (L - h) \rho_a} = \frac{\frac{1}{2}\rho_\ell h^2 + \frac{1}{2}\rho_a(L^2 - h^2)}{\rho_\ell h + \rho_a(L - h)}.$$

We now have the center of gravity, $\overline{x}$, as a function of the soda level, h. What value of h minimizes $\overline{x}$? We can compute $\overline{x}'(h)$ and set it equal to zero. This leads to the quadratic

$$h^2 + \frac{2\rho_a L}{\rho_\ell - \rho_a} h - \frac{\rho_a L^2}{\rho_\ell - \rho_a} = 0.$$

The positive root that corresponds to the physical solution is

$$h^* = \frac{\rho_a L}{\rho_\ell - \rho_a} \left(\sqrt{\frac{\rho_\ell}{\rho_a}} - 1 \right).$$

Using the given values of the constants ($\rho_a = 0.001$, $\rho_\ell = 1$, $L = 12$), we find that $h^* \doteq 0.368$. (When $\rho_a \ll \rho_\ell$, we can make the good approximation $h^* \doteq L\sqrt{\rho_a/\rho_\ell} = 0.379$.)

11.5 Let J be the amount of the jackpot, f the fraction of the jackpot awarded with the lump sum option, p the interest rate, I the annual withdrawal, and n the number of years that the annuity is paid. Let L_n and A_n be the balance in the lottery account after n years with the lump sum and annuity plans, respectively. Note that $\frac{J}{n}$ is the annual annuity payment. The difference equation for the lump sum option (assuming that the first withdrawal is taken immediately and that successive withdrawals are made after interest is compounded) is

$$L_{n+1} = (1+p)L_n - I \quad \text{with} \quad L_0 = fJ - I.$$

The difference equation for the annuity option is

$$A_{n+1} = (1+p)A_n + \left(\frac{J}{n} - I\right) \quad \text{with} \quad A_0 = \frac{J}{n} - I.$$

The solution of the lump sum equation is

$$L_n = fJ(1+p)^n - I\left(\frac{(1+p)^{n+1} - 1}{p}\right).$$

The solution of the annuity equation is

$$A_n = \left(\frac{J}{n} - I\right)\left(\frac{(1+p)^{n+1} - 1}{p}\right).$$

Using the given parameter values, we have the following balances: (i) With the lump sum option, the balance after 10 years is $8.3 million. With the annuity option, the balance after 10 years is $12.8 million. (ii) With the lump sum option, the balance after 30 years is $18.8 million. With the annuity option, the balance after 30 years is $16.5 million.

11.7 The goal is to find the function f such that $y = f(x)$ describes the path of the man. As the pursuit is taking place, both x and y are functions of t; in fact, $x'(t)$ and $y'(t)$ are the east-west and north-south components of the man's velocity, and $y'(x) = y'(t)/x'(t)$. The fact that the man's speed s is constant throughout the pursuit is given by

$$\text{Condition 1:} \quad (x'(t))^2 + (y'(t))^2 = s^2 \quad \text{or} \quad (x'(t))^2 \underbrace{\left(1 + \left(\frac{y'(t)}{x'(t)}\right)^2\right)}_{y'(x)} = s^2.$$

From this relation we find

$$\text{Condition 1a:} \quad x'(t) = -\frac{s}{\sqrt{1 + (y'(x))^2}},$$

where the minus sign insures that x is decreasing as it should. Noting that the position of the dog (moving at one mile per hour northward) is $x_d = 0$ and $y_d = t$, we can formulate the condition that the man always walks with the tangent to his path pointing toward the dog as

$$\text{Condition 2:} \quad y'(x) = \frac{y - t}{x} \quad \text{(which means } y'(x) < 0\text{)}.$$

Now the task is to reduce these two conditions in three variables to one condition in two variables by eliminating t. Differentiating Condition 2 with respect to t (*carefully!*) after multiplying across by x, we have

$$xy''(x)x'(t) + y'(x)x'(t) = y'(t) - 1.$$

Dividing through by $x'(t)$ and substituting for $x'(t)$ from Condition 1a, we have

$$xy''(x) = \frac{\sqrt{1 + (y'(x))^2}}{s}.$$

This equation is subject to the conditions $y(1) = y'(1) = 0$ (when $x = 1$, y and the slope of the pursuit curve are zero). The differential equation is second-order in y, but first-order in y'. Therefore let $v = y'$, and after separating variables, we have

$$\frac{dv}{\sqrt{1 + v^2}} = \frac{dx}{xs}.$$

Integrating once gives us

$$\ln(v + \sqrt{1 + v^2}) = \frac{\ln x}{s} + C \quad \text{or} \quad v + \sqrt{1 + v^2} = Cx^{1/s} \quad \text{for} \quad x \geq 0.$$

The condition $y'(1) = v(1) = 0$ implies that $C = 1$. Now solving for v leads us to

$$v = y'(x) = \frac{1}{2}(x^{1/s} - x^{-1/s}),$$

which can be integrated directly to give

$$y(x) = \frac{s}{2}\left(\frac{x^{\frac{s+1}{s}}}{s+1} - \frac{x^{\frac{s-1}{s}}}{s-1}\right) + C.$$

Using the condition $y(1) = 0$ tells us that $C = s/(s^2 - 1)$. The path of the man is given by

$$y = f(x) = \frac{s}{2}\left(\frac{x^{\frac{s+1}{s}}}{s+1} - \frac{x^{\frac{s-1}{s}}}{s-1}\right) + \frac{s}{s^2 - 1}.$$

We see that when $x = 0$, $y(0) = s/(s^2 - 1)$, which gives us the location of the encounter (in miles) of the man and the dog along the y-axis.

11.9 Letting $h(t)$ denote the depth of the water in the reservoir at time t and $\ell(t)$ denote the side length of the square formed by the surface of the water at time t, we see by similar triangles that $\ell(t) = 10 + \frac{h(t)}{2}$. Also the area of the surface of the water at time t is

$$S(t) = \ell(t)^2 = \left(10 + \frac{h}{2}\right)^2 = \frac{1}{4}(20 + h)^2.$$

Let $V_L = \frac{1}{3} \cdot 10^2 \cdot 20 = 666.7$ denote the volume of the (imaginary) pyramid below the water. Then the volume of water in the reservoir at time t is

$$V(t) = \frac{1}{3}S(t)(20 + h(t)) - V_L.$$

Noting from above that $(20 + h(t)) = 2\sqrt{S(t)}$, we can write

$$V(t) = \frac{1}{3}S(t) \cdot 2\sqrt{S(t)} - V_L = \frac{2}{3}S^{3/2} - V_L.$$

Thus,

$$S(t) = \left(\frac{3}{2}(V(t) + V_L)\right)^{2/3}.$$

Now we can use the assumption about the rate of change of the volume:

$$V'(t) = -\alpha S(t) = -\alpha \left(\frac{3}{2}(V(t) + V_L)\right)^{2/3} \equiv -a(V(t) + V_L)^{2/3},$$

where $a = \alpha(3/2)^{2/3}$. The volume can be determined with one integration, to give

$$V(t) = \left(C - \frac{1}{3}at\right)^3 - V_L,$$

where C is an arbitrary constant. Using $V(0) = V_0 = \frac{1}{3} \cdot 15^2 \cdot 30 - V_L \doteq 1583.3$, we find that $C = (V_0 + V_L)^{1/3} \doteq 13.10$. The volume function decreases monotonically and reaches zero at $t^* = 3(C - V_L)/a \doteq 199.6$ hours.

Bibliography

[1] B. Averbach and O. Chein, *Mathematics: Problem Solving Through Recreational Mathematics*, W.H. Freeman, San Francisco, 1980.

[2] R. Banks, *Towing Icebergs, Falling Dominoes, and Other Adventures in Applied Mathematics*, Princeton University Press, Princeton, NJ, 1998.

[3] R. Banks, *Slicing Pizza, Racing Turtles, and Further Adventures in Applied Mathematics*, Princeton University Press, Princeton, NJ, 1999.

[4] S. Barr, *Mathematical Brain Benders, Second Miscellany of Puzzles*, Dover, New York, 1982.

[5] H. Dudeney, *536 Puzzles and Curious Problems*, M. Gardner, ed., Charles Scribner's Sons, New York, 1967.

[6] W. Feller, *An Introduction to Probability Theory and Its Applications,* Volume 2, John Wiley and Sons, New York, 1971.

[7] R. Fraga, ed., *Resources for Calculus: Calculus Problems for a New Century*, Volume 2, MAA Notes 30, Mathematical Association of America, Washington, DC, 1993.

[8] M. Gardner, *The Scientific American Book of Mathematical Puzzles and Diversions*, Simon & Schuster, New York, 1959.

[9] M. Gardner, *The Second Scientific American Book of Mathematical Puzzles and Diversions*, Simon & Schuster, New York, 1961.

[10] M. Gardner, *aha! Insight*, Scientific American, New York, and W.H. Freeman, San Francisco, 1978.

[11] M. Gardner, *The Colossal Book of Mathematics: Classic Puzzles, Paradoxes, and Problems*, W.W. Norton, New York, 2001.

[12] G. Gilbert, M. Krusemeyer, and L. Larson, *The Wohascum County Problem Book*, Mathematical Association of American, Washington, DC, 1997.

[13] P. Halmos, *Problems for Mathematicians, Young and Old*, Mathematical Association of America, Washington, DC, 1991.

[14] P. Heafford, *The Math Entertainer*, Emerson Books, New York, 1974.

[15] R. Isaac, *The Pleasures of Probability*, Springer-Verlag, New York, 1995.

[16] M. Jackson and J. Ramsay, *Resources for Calculus: Problems for Student Discovery*, Volume 4, MAA Notes 30, Mathematical Association of America, Washington, DC, 1993.

[17] O. Jacoby and W. Benson, *Intriguing Mathematical Problems*, Dover, New York, 1996.

[18] A. Koestler, *The Act of Creation*, Dell Publishing, New York, 1964.

[19] J. Konhauser, D. Velleman, and S. Wagon, *Which Way Did the Bicycle Go?*, Mathematical Association of America, Washington, DC, 1996.

[20] S. Krantz, *Techniques of Problem Solving*, American Mathematical Society, Providence, RI, 1997.

[21] L. Larson, *Problem-Solving Through Problems*, Springer-Verlag, New York, 1983.

[22] S. Loyd, *Mathematical Puzzles of Sam Loyd*, Dover, New York, 1959.

[23] S. Loyd, *More Mathematical Puzzles of Sam Loyd*, Dover, New York, 1960.

[24] *The Contest Problem Book*, Volumes I–VI, Mathematical Association of America, Washington, DC.

[25] J. Madachy, *Madachy's Mathematical Recreations*, Dover, New York, 1966.

[26] S. Maurer, *Hat derivatives*, College Journal of Mathematics, 33, January 2002, pp. 32–37.

[27] P. Nahin, *Duelling Idiots and Other Probability Puzzlers*, Princeton University Press, Princeton, NJ, 2002.

[28] G. Pólya, *Mathematical Discovery: On Understanding, Learning, and Teaching Problem Solving*, John Wiley and Sons, New York, 1962.

[29] G. Pólya, *How To Solve It*, 2nd ed., Princeton University Press, Princeton, NJ, 1971.

[30] C. Quigg, *Mathematical Quickies*, Dover, New York, 1985.

[31] P. Renz, *Thoughts on innumeracy: Mathematics versus the world*, American Mathematical Monthly, 100, 1993, pp. 732–742.

[32] S. Ross, *A First Course in Probability*, 6th ed., Prentice–Hall, Englewood Cliffs, NJ, 2001.

[33] M. Rubinstein and K. Pfeiffer, *Concepts in Problem Solving*, Prentice–Hall, Englewood Cliffs, NJ, 1980.

[34] M. Rubinstein and I. Firstenberg, *Patterns of Problem Solving*, Prentice–Hall, Englewood Cliffs, NJ, 1995.

[35] A. Schoenfeld, *Mathematical Problem Solving*, Academic Press, New York, 1985.

[36] P. Sloane and D. MacHale, *Challenging Lateral Thinking Puzzles*, Sterling Publishing, New York, 1993.

[37] P. Sloane and D. MacHale, *Great Lateral Thinking Puzzles*, Sterling Publishing, New York, 1994.

[38] M. Spiegel, *Schaum's Outlines: Probability and Statistics*, McGraw–Hill, New York, 1996.

[39] T. Stickels, *Are You As Smart As You Think?*, Thomas Dunne Books, New York, 2000.

[40] P. Vaderlind, R. Guy, and L. Larson, *The Inquisitive Problem Solver*, Mathematical Association of America, Washington, DC, 2002.

[41] W. Wickelgren, *How to Solve Mathematical Problems*, Dover, New York, 1974.

[42] P. Winkler, *Mathematical Puzzles, The Connoisseur's Collection*, A K Peters, Wellesley, MA, 2004.

[43] P. Zeitz, *The Art and Craft of Problem Solving*, John Wiley and Sons, New York, 1999.

Index